Aunt Carrie's War Against

BLACK FOX

NUCLEAR POWER PLANT

Aunt Carrie's War Against
BLACK FOX
NUCLEAR POWER PLANT

CARRIE BAREFOOT DICKERSON

with Patricia Lemon

P.O. Box 2584
Claremore, OK
74018
918-341-2919

To Jeanette Lewis
With best wishes
Carrie Barefoot Dickerson
7-16-2002

Council Oak Publishing   Tulsa, Oklahoma

Council Oak Publishing Company, Inc.

Tulsa, OK 74120

©1995 Carrie Barefoot Dickerson. All rights reserved

99 98 97 96 95   5 4 3 2 1

ISBN 1-57178-009-2

Book and cover design by Carol Stanton

Library of Congress Cataloging-in-Publication Data
Dickerson, Carrie Barefoot, 1917-
        Aunt Carrie's war against Black Fox nuclear power plant / Carrie
Barefoot Dickerson with Patricia Lemon.
              p.   cm.
        Includes bibliographical references and index.
        ISBN 1-57178-009-2 (cloth)
        1. Nuclear power plants--Oklahoma--Rogers County--Planning.
2. Nuclear power plants--Oklahoma--Rogers County--Public opinion.
3. Antinuclear movement--Oklahoma--Rogers County.   I. Lemon, Patricia.   II.
Title
TK1344.05D53   1995
333.792'415'0976694--dc20   95-17820
                                                     CIP

In memory of my husband, Robert

Table of Contents

Acknowledgments

I would like to be able to thank thousands of people, individually, for their help with this book. Every member of CASE, the Sierra Club, and the Sunbelt Alliance has contributed in some way, as have many Public Service Company of Oklahoma (PSO) employees—not to mention dozens of people in federal, state, and local government and anti-nuclear activists across the country. Many of you are mentioned in these pages by name. Many, many more are anonymous. Your moral and material support have made all the difference to me. You all know who you are, and I know you will forgive me for not trying to list all your names here.

This book has been long in the writing, and without continuing encouragement over the years from Dan Agent, Nadine Barton, Susan Farrell, Cathy Coulson, Dennis Farrell, Helen and Barbara Geary, Doris Gunn, Betty Knight Broach, Theresa Miller, Peggy Whitt, and John Zelnick, it might never have even seen the light of day. When my heart sank within me at the thought of reviewing all that had happened and trying to make sense of it for others, they told me how important it is for everyone to know what we still face and how important it is that we all know we can control our own fates, however incompletely. When I thought I would never be able to write anything readable, they reminded me that I had been making people listen for years. When I thought I would never be able to find the energy to finish, their love and encouragement gave me strength.

By the time I had realized that I really could write the book, Janet Hutto introduced me to Sally Dennison and Paulette Millichap who offered me a publisher that I felt I could trust to let me tell the story without distortions other than those of time and my perceptions and memory.

My sister, Florence Cahalen, read the manuscript at an early, very crude stage and made suggestions that have helped me enormously, and I am grateful for the comments of another sister, Paula Bentley, on a number of family stories. Frequent calls from my twin, Clara Barefoot-Sehorn, have kept me going in the most difficult times and given me ideas that have added immeasurably to everything I have written.

At various stages in the process, Mildred Ayers, Eddie Bryant, Robert Butler, Kathy Dalrymple, Joyce Nipper, Nancy Perreault, and Marjorie Spees have read and commented helpfully on the manuscript, as have my daughter-in-law Kay Dickerson, my grandchildren Ted Lemon and Sandra Patterson, my cousins Marilyn and Dick Bell, and Robert's cousins Ruth Anne and Clem Hutchinson. Irene Dickinson, Nancy Dodson, Kay Drey, and Mary Olson have been constant sources of up-to-date information about the anti-nuclear movement nationwide, and Dr. Benjamin Spock and his wife, Mary Morgan, have continued to encourage me in my work.

My son-in-law, Edward Lemon, not only stepped in to help my daughter Patricia make sure I would continue to have a roof over my head while I wrote, he also found the right computer and word-processing programs to make my task simpler, and Martha Kitchen donated her printer to the cause.

Without Tom Bomer's presence of mind, there would have been no records for me to consult in the writing. When our lawyers concluded their work and we closed the CASE office, Tom retrieved and brought to me truckloads of documents; they filled two rooms of my home, floor to ceiling, and there they remained for years. When the time finally came, I had the information I needed—because of Tom.

Above all, however, the successful completion of this book depended on the hard work of two of my daughters, Mary Dickerson and Patricia Lemon. Patricia spent weeks sorting out what Tom had retrieved. My nephews Michael Cahalen and Rod Sehorn, discarded the duplicates because office paper was not recycled then. Mary, who has been my support in the fight against Black Fox and my constant companion since Robert's death, has spent uncounted hours of her life finding and checking out sources for me and entering my writings into the computer.

While all this was going on, my daughter Florence Snelling fed our spirits with flowers, and my son Jim fed our bodies with mushrooms and pecan pies. Sandra's husband David made sure that Robert's peach orchard would continue to delight me every spring, and their son Wesley provided the joys of great-grandmotherhood. J.J. Dickerson and Melissa and Kevin Price looked in on me every now and then; Ted and Signe and

Signe's husband Paul Friedrichs came to see me from what seems like the ends of the earth, making me feel that they understood and forgave me for neglecting my grandchildren in my preoccupation.

In the end, it was Patricia who carefully weeded out the unnecessary verbiage from the manuscript and helped me add the little stories I had been telling her for years but had never thought to put in the book.

All of you have my profound gratitude. All of you are in my thoughts and prayers.

Introduction

I was ambivalent about writing this story, my memoir of a successful campaign to prevent construction of two nuclear power reactors. My first inclination was to let sleeping dogs—or rather, the dead Black Fox nuclear plant—lie. There were many painful memories that I was reluctant to dredge up again, many unhealed scars that I did not want to touch. At the same time, I knew that unless I found the courage to reexamine them, they would never heal. Even worse, unless I found the courage to tell people throughout the country that democracy is not dead and that we can shape our own political lives, I would have left undone a duty to my fellow citizens.

Recent political activity in the United States has seemed to be a reflection not of what we want, but simply of what we don't want. It is not necessary to destroy our economy to prevent big business from blighting our lives and those of our unimaginably distant descendants. If we choose only to destroy what attacks our lives without putting in place that which will protect and enhance them, we will have destroyed ourselves just as surely as we defeat what we are fighting.

The nine-year battle to stop the Black Fox facility, beginning May 8, 1973, was, at times, too painful to think of recounting. It would have been easier to put that chapter behind me and to get on with living my life. Never in my wildest dreams did I envision being cast in the role of activist, protester, and intervenor against nuclear power. The years of hard, unrelenting work, the unavoidable confrontations, and the suspense we endured through the long waits for final court decisions, left me physically and emotionally exhausted and financially impoverished. I dreaded reliving those years.

Yet overriding my personal feelings was a feeling of obligation to society. I have a responsibility to the people who helped me stop Black Fox, to the victims of Three-Mile Island and Chernobyl, to the victims of other nuclear tragedies, including those in our own state of Oklahoma, and to potential victims of possible future nuclear disasters. Because of the widespread effects of the Chernobyl disaster, we now know that the problems of nuclear power can affect everyone, that there is no safe place. There are no fence lines, no boundaries, no safeguards, to contain radioactive fallout.

Recent revelations of problems in nuclear power plants in the U.S. have shown us that we cannot afford to be complacent. While our plants have different vulnerabilities, they are as susceptible to disaster as Chernobyl. While the mining of uranium continues, every person everywhere in the world is at risk from airborne radiation and, thus, from radiation-induced genetic damage. Even if the mining were to stop tomorrow, we would continue to be exposed to the residue from past actions. While radioactive waste continues to be generated, we remain susceptible to atomic detonations generated when a critical mass is accumulated unwittingly.

We are now increasingly aware that the burning of fossil fuels produces carbon dioxide and other gases that are partially responsible for the greenhouse effect which has begun to affect the climates of the earth. Scientists warn that global warming may cause the polar ice caps to melt, raising the levels of the earth's oceans and inundating coastal and low-lying lands of the continental interiors. Even now, inhabitants of some cities in developing countries must rely on the meager protection of masks to preserve their lives.

"Nuclear power," say its proponents, "is the energy solution for the future; it does not produce carbon dioxide." Many who have come to fear nuclear power at home still advocate its use in developing countries on the grounds that distance will protect us. Both are wrong. True, nuclear power does not produce carbon dioxide; instead, it produces radioactive waste. Carbon dioxide may change the face of the planet and shorten lives now, but radioactive waste is capable of destroying all life on earth, now and for all time. We must not trade one terrible evil for a worse one.

Instead, we must radically improve energy efficiency and conservation and adopt renewable energy sources—sun, wind, water and hydrogen—a course of action that is perfectly feasible and that environmentalists have advocated for many years.

The framers of our federal constitution provided avenues for citizens to make their voices heard and for the redress of wrongs. When government becomes unresponsive to citizen concerns, there is always the temptation to tell ourselves that the system has failed and the only solution lies in direct action. At times that is true, and at the time I am writing about, many people believed it to be true in this country. Unfortunately, direct action is a two-edged sword; because those who use it are convinced they are right, they may not allow for public discourse. In the hands of a Ghandi or a Martin Luther King or most of the members of the Sunbelt Alliance, it is benign. In the hands of a Hitler or an Ayatollah Khomeini or one of the American demagogues currently encouraging murder in the cause of preserving life, it often destroys nations. With the best of intentions, direct action, too, can become an example of trading one terrible evil for a worse one.

In common with a number of other organizations, the group that I founded, Citizens' Action for Safe Energy (CASE), demonstrated that the avenues our forefathers put into the Constitution are still open and available and that, for all the doomsaying we read daily in the newspaper and hear in the broadcast media, democracy is still alive and still worth preserving. Through intervention in court proceedings, we fought city hall—the federal Nuclear Regulatory Commission, the industrial giant General Electric, our local power company, and the legislators seduced by their various lobbying organizations—and won.

I believe that it is even more urgent now than it was all those years ago that citizens of this country be aware that we have viable alternatives to nuclear power. I believe it is even more urgent now that we understand the true dimensions of the nuclear threat. I also believe it is important that we regain our faith in our ability to make democracy work for the good of all. If, in relating the saga of Black Fox, I succeed in contributing to this goal, my efforts will have been worthwhile.

—Carrie Barefoot Dickerson
Claremore, Oklahoma
May, 1995

Note: I dislike endnotes, because they make me think I have to read them, so I have kept them to a minimum in this narrative; read them only if you need to know a specific source. The glossary is there for acronyms and words I didn't know before Black Fox.

WORLD

RELIABILITY
CHARACTER
ENTERPRISE

Single Copy—10c

40 PAGES—2 PARTS

DAY, MAY 8, 1973

$450 Million N-Plant Proposed by PSC For Site Near Inola

Public Service Co.'s proposed $450 million nuclear power plant will be turning out 1,100,000 kilowatts by 1983 at a site southwest of Inola — if an environmental impact study is approved.

A PSC spokesman said

ing engineers, compiled a preliminary study which said the quantity and quality of water available are suitab'e and that natu-

water
am.

Sargent and Lundy reported that the level of dissolved minerals and salts is to prevent prema and corrosion of cooling system

Morph

study will include geological aspects, including soil and bedrock structure, topography and seismic activity.

"In addition to meeting ificat.ons fr
on A-4

N-Power Plant Is Proposed

Continued from A-1

a nuclear plant, this site does not infringe on special areas that must be preserved," Morphis said.

"It is isolated from areas of high population density, parks, historical sites, hunting grounds and oil fields.

"We took care to avoid unspoiled wildlife areas that could be jeopardized by industrial development. We're also far enough away from military areas and heavily traveled airspaces that could pose potential hazards to the facil-

" he added.

N-Plant?

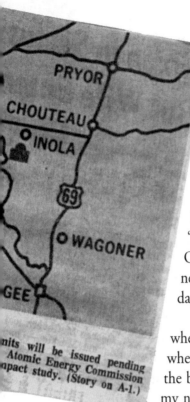

I still don't know who placed the newspaper on my office chair that spring morning, May 8, 1973. It had been folded so the headline was clearly visible: "$450 M N-Plant Planned for Inola." N-Plant? I discovered that "N" stood for nuclear: Public Service Company of Oklahoma (PSO) planned to build a nuclear power plant near Inola, a Rogers County town not far from us. That day, quietly and without fanfare, my life was changed.

It was my custom in good weather to pedal my three-wheeler the three miles to work, and that morning as my wheels hummed along, I had been a free spirit, reveling in the beauty of the early May morning. I arrived prepared for my normal, busy day supervising the care of the residents of Aunt Carrie's Nursing Home in Claremore. My work in the hundred-bed home my husband and I had built in 1964 filled my life. I had no need for other responsibilities.

After they finished their farm chores, my husband Robert and our daughter Mary would always follow me into town. Robert was responsible for administration, and Mary supervised the kitchen. After everyone else had finished breakfast, the three of us would sit down to ours. That was our time of togetherness, of sharing events, and of planning the schedule for the day.

That day, my early morning routine completed, I returned to the office for a much needed break—only to find the newspaper lying in wait. The weight of the world descended on my shoulders that morning—and I bore its full measure for nine long years. Never since have I been able to recapture the carefree exhilaration of that morning, as I pedaled my bike to work with the rush of the wind in my face.

Working such long hours, I rarely had time then to read newspapers, and almost never listened to the radio or watched television. If that newspaper had not been left for me to read, the nuclear power plant could have been built and operating before I even knew about it.

Somewhere in my memory was a 1950s magazine picture of a muskrat, with a caption that read something like this: "Radioactive effluent, released into the tributaries below Oak Ridge Nuclear Facility in Tennessee, gave this muskrat cancer." At the time I read it, I must have been very naïve, because I remember thinking, *I am sure our government will take care of the problem; now that they know what it can do, they will stop the release of radiation.* But had they? I didn't know. If radiation would give cancer to a muskrat, what would it do to me and my family?

I consulted a Rogers County section map and discovered that the site of the proposed plant was twelve miles due south of our farm. If the government had not, in fact, taken care of the Oak Ridge muskrat problem and if radiation from nuclear facilities was still causing cancer, twelve miles was too close.

I knew nothing about nuclear power, so I phoned the Atomic Energy Commission and asked them to send me everything an ordinary citizen could understand about the subject. For months after that telephone call, a large packet arrived every week from the AEC.

When I was growing up, my family taught me that the Bible enjoined us all to be responsible stewards of the Earth, and in 1962, Rachel Carson's book, *Silent Spring*, awakened me to an awareness of modern environmental problems. For several years after that, I had kept an environmental scrapbook, and in it I found the address of the Sierra Club in San Francisco; they told me that there was an Oklahoma chapter. I joined, hoping that the Sierra Club would take as one of its projects, opposition to Black Fox (the name later given the proposed plant). However, it was not until the first week of 1974 that the national Sierra Club board voted nine to four in favor of an anti-nuclear policy. At about the same time, I became a member of the state board of the Oklahoma chapter, and eventually the Sierra Club became a valuable ally in the fight to avert the nuclear facility at Inola.

I also wrote to an organization called Friends of the Earth (FOE). They quickly recruited me into their national network of anti-nuclear activists, of whose existence I had been completely unaware. I had discovered a whole new world!

At the time of the nuclear power plant announcement, my supervisory duties and my work as the registered nurse for our nursing home kept me busy fifteen to eighteen hours every day. I hired a nurse to take over one eight-hour shift of my work, and each day after work I went home and read and researched the information I was receiving from many sources. For three months I studied eight, ten and sometimes twelve hours a day. I was often too engrossed in the information to sleep. The more I read, the more concerned I became.

I learned the disquieting fact that Oklahoma had already been invaded by nuclear facilities. On their campuses, Oklahoma State University and the University of Oklahoma each had a small research reactor where nuclear engineering was taught. (They have since been removed from both campuses.) But those were just the tip of the iceberg: The Kerr-McGee Company, which I had thought of only as an oil company, had been operating nuclear fuel processing facilities in Oklahoma at three locations, two at Crescent, twenty miles north of Oklahoma City (which were closed in 1975), and one in the far eastern part of the state at Gore (which, with a second built in 1986, closed in 1993). A thorium processing plant Kerr-McGee had operated at Cushing had exploded, killing one person and polluting a wide area with radioactive materials. That plant had been decommissioned in 1966 and its license terminated—too late for its victims.

In 1938, two European scientists succeeded in splitting an atom of uranium. They and others who dreamed of expanding this "fission" into a chain reaction fled Nazi Germany and emigrated to the U.S. and Canada. Three of those scientists persuaded Albert Einstein to ask President Roosevelt to form a three-member "Uranium Committee." On August 13, 1942, began three incredible, fast-paced years during which, at a cost of about two million dollars, the top-secret Manhattan Project developed the atomic bombs that destroyed Hiroshima and Nagasaki on August 6 and 9, 1945.

To carry out this work, three huge facilities were built, one at Oak Ridge, Tennessee, which concentrated uranium, one at Hanford, Washington, where plutonium was produced, and one at Los Alamos, New Mexico, where the bombs were assembled. Government records indicate that before the bombs were dropped, Japan was seeking a peace settlement. Those scientists who had initially urged the development of the bombs eventually asked Secretary of War Henry Stimson not to use them against Japan. At that critical point President Roosevelt died. His successor, Harry Truman, gave the go-ahead for the first bombing, apparently believing that we would otherwise have to invade the Japanese mainland and thereby incur thousands of American casualties.

President Eisenhower's Atoms for Peace Program, some feel, was designed to assuage our nation's feelings of guilt for the suffering caused by the atom bombs dropped on Japan and to give Americans a financial stake in continued nuclear development and experimentation. Private industry was invited to profit financially from a process that had been developed with public funds by public workers. One of the most short-sighted provisions of the program made the Atomic Energy Commission responsible both for promoting nuclear energy and for protecting the public from it.

The private electric utilities hesitated, fearing nuclear power would be too expensive to produce and that, since insurance companies refused to insure the plants, they would be left holding the bag if the plants proved unsafe. Following its mandate to promote nuclear power, the AEC issued a veiled warning to the electrical industry that if private utilities balked, the government might set itself up in competition. The accompanying carrot was a proposal to build demonstration reactors with taxpayers' money and to supply fuel at attractive prices through government subsidies.

I soon learned that insurance companies had good reason to refuse to insure nuclear power plants. Brookhaven Laboratories' 1957 report to Congress on the hypothetical consequences of a nuclear accident (contained in its 1956 study, *Wash 740*) predicted that within fifteen miles of a small reactor meltdown, 3,400 people would be killed and that within forty-five miles, 43,000 people could be injured. Furthermore, it suggested, property damage could reach $7 billion, and an area the size of Maryland could be contaminated.

Instead of drawing the obvious conclusion—that the nation and the world could not afford the potential damage of nuclear power—Congress passed the Price-Anderson Act of 1957, limiting total liability for a nuclear accident to $560 million and utility company liability (which would be covered by insurance) to only $60 million, with taxpayers covering the rest. Critics pointed out that victims might collect only eight cents on the dollar, but the American Nuclear Society exulted that the measure was a "real vote of confidence," because the $60 million was the "greatest commitment [insurance companies] have ever made for a single hazard."

In 1965, *Wash 740* was updated to include the larger reactors developed since the 1957 report. The second report predicted that if the Emergency Core Cooling System failed to work when water to the nuclear power plant was cut off, the nuclear fuel could become a molten mass that would burrow deep into the earth ("all the way to China," as we used to say when we were children; hence, the China Syndrome).

During such a meltdown, radioactive materials would become gaseous and escape into the atmosphere, causing fallout equivalent to the release of several thousand Hiroshima-sized bombs. The fallout would blanket an area the size of Pennsylvania, rendering it uninhabitable for centuries. It was predicted there would be 45,000 deaths, 100,000 injuries, and $18 to $280 billion (in 1965 dollars) in property damages. Many thousands of people would die of acute radiation poisoning and thousands more of lingering radiation sickness, whose symptoms include bleeding gums and mucous membranes, vomiting, diarrhea, and severe headaches. Farther from the meltdown site, infants would be born with severe deformities and retardation, children and adults would develop leukemia and other forms of cancer, and unborn generations would be genetically damaged.

It was very nearly a description of the Chernobyl disaster on April 26, 1986.

The electric utilities were convinced that if the public were ever to learn of Brookhaven's predictions, the nuclear power business would be crippled or perhaps even killed. So for eight years, the AEC kept the study secret—choosing promotion over protecting the public.

In early 1973, several individuals and groups, including environmental attorney Mike Cherry and activist David Dinsmore, both from Chicago, Ralph Nader, and the Union of Concerned Scientists learned about the report. They threatened to file a Freedom of Information Act suit. As a result the report was made public about the time I phoned the AEC. One of those weekly packages contained a copy. All I could think was, *Oklahoma is about the size of Pennsylvania. Oklahoma would be eradicated by a power plant accident. We cannot allow that threat to hang over our heads.*

The AEC came under a tempest of criticism for its concealment of the report, and to defend its stand, it commissioned a new study from a group headed by MIT nuclear physicist Norman Rasmussen. Rasmussen's study, *Wash 1400*, avoided the worst-case approach in favor of a probability and likely-consequences approach. The report calculated probability of a nuclear accident as one in 10,000 per year per 100 reactors, or one in a million a year for a community near a specific atomic plant. They said that the chances of a nuclear accident would be like "a meteorite striking the New York Metropolitan Life building."

In other words, the group held that if there were 100 reactors in the world, there would, on the average, be one accident every 10,000 years—or if you lived in a community near an atomic plant, you could expect to be killed once in a million years. If one accepts the premises of the calculations, this sounds comforting. Unfortunately, however, there is nothing in such predictions to say in which year the accident would occur; it could just as easily be the first as the last.

In the book he wrote with his colleague Arthur Tamplin, *Poisoned Power: The Case Against Nuclear Power Plants*, John Gofman claimed that a runaway reactor could kill *5 million* people, not 10 thousand. In 1969, convinced that nuclear plants threatened to create a public health tragedy, he took the lead in the struggle against their proliferation. Gofman, M.D., Ph.D., professor *emeritus* of Medical Physics at the University of California, Berkeley, is one of the experts who disagreed with the Rasmussen report. In 1942, Gofman and Tamplin isolated the world's first milligram of plutonium. Gofman, who received the Stouffer Prize for his research in lipoproteins and heart disease, had been Associate Director of the Lawrence Livermore (radiation) Laboratory, where he conducted research on cancer and chromosomes—until the Atomic Energy Commission stopped his work in 1972.

While Gofman was one of the most vocal of Rasmussen's critics, he was far from alone. The report has been so widely discredited on conceptual and mathematical grounds—and by the events of intervening years—that by 1994 it is pretty much a dead letter. For years, however, proponents of the nuclear industry would point to it proudly whenever their claims were brought into question.

"Nothing can stand
in the way ..."

Inola is a small town in the midst of a quiet farming community about twenty-five miles east of Tulsa. Amish, Mennonite, and German farmers settled on the land decades ago. The "Hay Capital of the World," Inola is a community of hard working, God-fearing people, a place where anyone might wish to live. I read in some of my Sierra Club and FOE publications that utilities tend to place their nuclear power plants in just such settings— in small communities where people are quiet and occupied with their own affairs. Such communities are less likely to interfere in what they see as others' business.

*Carrie with daughters
Patricia & Florence*

Inola Builders Black Fox Boo

**Story and Photos
By PAUL HART
World Real Estate Writer**

LA — "Drastic changes" are coming for this town which spro siding of the Missouri Pacific Railroad in 1889. Inola has alrea from a farm town into a Tulsa bedroom community — and soo me a city in its own right.

quite a bit of building going on but it's not a boom — yet,"
Riggs, who has been building houses here for eight years.
e years there will be drastic changes," said Don Allen, a
iction worker who started his own contracting firm 1½
he saw a growing demand for new housing.
vasn't too far back he was building six or seven ho
at demand is from Tulsa workers who want
outting up 18 or 19 houses each year.
cording to Riggs. State Highway 33 n
five miles between here and
soon on four-laning the ro
69 near Chouteau.

THE Ino
'Nuclear' L

By Jim Etter

INOLA—The first nucle-
ar electricity-producing
plant in Oklahoma and its
1,900,000 kilowatts apparently will be welcome in
this small "Hay Capital of
the World" community.

Many in the Rogers
County town of about 1,500
are looking forward to between 500 and 1,000 jobs
during construction—even
though the work won't
start before late 1977.
Others are proud
ola facility
early

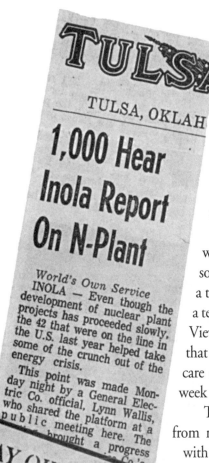

In the early 40s, I had spent my first year of teaching in a two-room schoolhouse in the German-speaking Mennonite and Amish community of Pleasant View, only three miles southeast of Inola. The students spoke German at home and on the playground. With them and their families, I attended the German-language services in the Mennonite churches of the community. I have always regretted that I had no time to take advantage of that wonderful opportunity to learn to speak a second language.

We had had a crop failure on the farm, and we would have been starving and destitute without some other source of income. I had been taught to be a teacher, so it seemed only natural for me to look for a teaching position. The only one open was at Pleasant View, and it was with mingled anticipation and regret that I prepared to leave my two little daughters in the care of my mother-in-law while I boarded during the week in homes in the community.

That was the hardest part of that year, being away from my little girls. My mother-in-law was very good with them, but when I came home on weekends, my baby Florence, who was not yet two years old, followed me around holding onto my little finger. My heart would break every time she asked, "Mommy, why do you leave your little girl? Don't you love your little girl?" How could I make her understand that gas rationing made it impossible for Robert to drive me to the school and back each day?

With the children where I boarded, I rode to the school each morning in a horse-drawn buggy and taught students from the first through eighth grades. Each morning, I built a coal fire in the pot-bellied stove before the other children arrived. There were no modern conveniences, and the bathroom was an outhouse at the edge of the school yard. At the end of each school day, the children and I swept the floor and tidied the schoolroom and locked the door behind us.

I prepared lunch every day for the students—usually a pot of stew or soup of some kind—atop the pot-bellied stove. Occasionally I would also bake cornbread in a skillet on top of the stove. Since there was no refrigeration, our fare was simple, but we did have carrot sticks and fresh fruit in season. Some of the food was supplied by a school lunch program, and on weekends I shopped for the rest.

One day each week, Mabel Rucker Jones came to the school and gave each of the fourteen children a half-hour music lesson. All my life I had dreamed of playing the piano, but we had been too poor to pay for lessons, much less a piano. Now, dear Mabel took her lunch hour to give me a music lesson each week. That year, I finally learned to play the piano for my own enjoyment. My first paycheck went to pay for a flock of sheep for the farm. My second paid for an old upright Steinway with a beautiful tone, one of the best investments I have ever made. For years, it was the center of our life, and now it sits in my granddaughter's house.

Mabel started coming to my home on Saturdays to teach my four-year old, Patsy, awakening her interest in music. She loved to sing, but it took years for her to learn to enjoy practicing the piano. When I would tell her about looking at her long fingers when she was born and saying to myself that they were pianist's fingers, she would retort, "But I don't want to be a concert pianist, I want to sing." Sing she did, winning prizes in state contests and going on to participate in Choral Society and Gilbert and Sullivan in college and Blanche Moyse's Marlboro Music

Festivals years later—all sparked by my first investment. She married a scholar who is also a professional pianist, and both her children are deeply musical. The other children enjoyed music, but none took advantage of the piano the way Patricia did.

Twenty years later I returned to Inola, this time to teach home economics at Inola High School. Over the years, I developed warm friendships with Inola people and accumulated wonderful memories of my teaching experiences there. Even Inola's mayor, Tommy Dyer, had once been a favorite fifth-grade student of mine in the little community of Tiawah.

PSO bought most of the 2,000-acre plant site through a third party to keep land prices from climbing steeply and avoid attracting adverse publicity before they were ready to make their announcement. Once they had made their plans public, however, PSO threatened to use eminent domain proceedings to force those who had preferred to keep their land to sell. Some landowners, including the Beaver family, brought lawsuits against PSO, forcing them to make fair market payments for the land.

To convince Inola residents of the wonders of nuclear power, PSO invited the community to a get-together and progress report. Despite a drenching rainstorm, the school gymnasium was packed. PSO brought a thousand chicken dinners and a live band. After wining and dining them, PSO proceeded to sell the people a bill of goods: The citizens of Inola were told that the enormous taxes the plant would bring would help them build the best school in the State of Oklahoma and were assured that nuclear power was the cheapest, cleanest, and safest available source of energy.

I wanted to see the land where PSO planned to build the plant, so in late May I drove through their open gate about three miles southwest of Inola. I was the lone trespasser. Not another soul was in sight as I drove along the winding lane and across a little bridge into a woodland. In its breathtaking beauty it was a fairyland. Overlooking a rolling meadow where wild flowers swayed in the breeze, majestic trees several feet in diameter—trees that were surely over two hundred years old—guarded its peace and tranquillity. Wildlife abounded. A rabbit hopped across the lane, and squirrels scurried in the trees. Bird songs filled the air, while a red-tailed hawk soared above the meadow, searching for prey. I gave the right of way to a covey of quail crossing the lane, the fluffy, brown chicks following single file behind their anxious mother. This Eden—to be uprooted?

Nuclear power plant proponents comfort people who mourn the passing of such beauty by reminding them that AEC law requires the utility company to restore the land to its original state if a construction permit is denied. Before we were able to halt construction at the site, PSO was allowed to change the face of the earth in preparation for the building process. How could they replace 200-year-old trees? How could they reconstruct bulldozed terrain?

I telephoned PSO president R.O. Newman and asked just those questions. "Is the issuance of the construction permit cut and dried?" I concluded.

"Nothing can stand in the way of the completion of the nuclear power plant," was his daunting reply.

I refused to allow myself to visualize the completed facility. I knew I must hold in my mind the reality of the natural beauty I had seen before PSO started moving dirt and changing the shape of the land, because that is the only place it still exists.

Carrie's mother, Carrie Barefoot and brother Marvin.

Love thy Neighbor

The first weeks of study and research about nuclear power left me angry and weak with horror. How could PSO people, people who lived among us and shared the same surroundings, wish to build a plant with such disastrous potential for people, property, and the environment? How could my government, the people I depended upon to represent me and to look after my well-being, so betray us all? My waking nightmare invaded my dreams and kept me from sleeping. Soon I was so depressed and ill that it was a struggle for me to carry out daily tasks and speak cheerfully with people at the nursing home.

I knew that I faced a terrible conflict, and I knew that to win it I must be healthy and strong. Something had to change. I

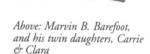

Above: Marvin B. Barefoot, and his twin daughters, Carrie & Clara

Left: Grandmother Nancy Clara Withrow Barefoot

began to see that my anger was unproductive—I was the person it hurt, not the misguided officials of PSO and the government. I knew that illness and grudge-bearing would defeat my efforts.

In my distress, I reached back to the years of my childhood and youth in the 20s and early 30s. Long-ago memories, half-forgotten in my busy life, surfaced, and I could hear the voice of my widowed mother admonishing us in the words of the Bible to "love thy neighbor as thyself."

I recalled my first and most cherished memory, my only memory of my father. He was sitting at our dining room table in a blue room, holding my twin sister and me in his arms. I remembered my happiness at being with him—I had felt so loved.

We lived in the country thirteen miles west of Okmulgee, where Clara and I were born on May 24, 1917, in a log cabin—an hour before the doctor arrived to cut our umbilical cords. I was born with an umbilical hernia (which was corrected years later), and both Clara and I were born with yellow jaundice. The doctor laid me at the foot of the bed and told our mother I wouldn't live but that Clara was robust and healthy. "Give your attention to the redhead," he said to her. "Don't waste your time on the other one." I guess I fooled him!

We moved from the log cabin to the prairie, and there our father rented land and borrowed money to start a cattle ranch. (He met my mother when he was buying cattle from her. The beautiful widow had saddled her pony and helped him round them up.) He had several teams of horses and hired hands who helped with the ranch.

To make extra money, he installed a large water tank on a wagon, and one of his hired hands supplied water to the oil field workers in the nearby Youngstown oil field. Many years later, that same man came to live in our nursing home. When Charles Flanagan learned my maiden name, he began telling me stories of my father and of his rounds on the water route and how beautiful and strong our horses were. He told of carrying Clara and me in his arms when we were babies. It was a joy to me to recover that piece of my father's life, and I believe it was an equal joy to him to have an opportunity to relive his youth.

By early 1920, our mother was again expecting a baby. Because our father didn't want her to repeat the same frightful experience as Clara's and my birth, he moved us into Okmulgee for the birth of the baby. But before the new baby had time to arrive, our father had contracted the German flu. We went back to the ranch where my mother could care for him until she, too, fell ill. Mama almost died. Daddy did. I don't remember anything about his death on February 13, but Clara says she remembers crying and running after the hearse as it drove away with his body. Grandmother Perry had to run after her to bring her back.

To pay off the bank loans, the cattle and all our belongings were sold, but we still owned the eighty acres on Salt Creek at Rocky Hill. Our mother's first husband had left it to our older brother, and before our father's death, our mother's older brother, Isaac, had built a four-room frame house there.

Grandmother Perry came to live with us when our father died and accompanied us when we moved back to Okmulgee for the birth of our little brother. Marvin was born at ten o'clock, the morning of February 23, 1920. And that was my second memory!

Children were not supposed to witness the birth of a baby in those days, but the doctors were so busy taking care of my mother that they didn't even notice me standing in the doorway watching the event. We must have lived close to school, because my nine year old brother, Joe, came home for lunch every day. I remember watching impatiently for him to come down the street that day. The instant he came in sight, I ran out to tell him the wonderful news, screaming excitedly, "We have a new little baby brother!"

I remember, too, my terrible fall. We went shopping for a baby buggy for Marvin. The department store was so new that the railings were not yet in place around the mezzanine floor where we found the baby buggies. A curious not-yet-three-year-old, I wondered what was beyond the edge of the floor. Too quickly for grown-up reflexes, I pulled loose from my grandmother's hand, peered over the edge, and fell headlong onto the floor below. I remember the clerk reaching out to try to catch me, and that is the last of that memory.

Later I was told that I had been unconscious for three hours. The doctor kept telling my mother he could do no more, but my mother refused to allow him to stop working on me until I regained consciousness. I'm alive today because of her inability to take no for an answer, but I still suffer from the skeletal damage the fall did. Even so, I have never allowed my problems to keep me from working hard to achieve my goals.

I thought of the little three-room Rocky Hill Grade School (near the western shore of what is now Lake Okmulgee), that was built on three acres of our farm. There we learned the "three Rs" and there Mrs. Bunch, J.D. Campbell and Blaine Glover taught us to be honorable, upright, and responsible citizens. And I thought of Nuyaka High School (in western Okmulgee County), where Mr. Rainwater illuminated the state and federal constitutions for our benefit, Mr. Norvell led us through the intricacies of mathematics and probability, and Miss Russell heightened our interest in literature. I remembered community church services at the little grade-school building, where we sang beautiful old hymns, and Aunt Meda played the pump organ. I remembered singing conventions and community picnics.

The year I was ten, I had my first taste of sacrifice for the good of the community. After years of work, the Lake Okmulgee Dam was finally complete, and the rising lake was soon to cover the fertile bottomland of our farm. In a traumatic—but exciting—family upheaval, our four-room frame house was moved about three miles west, as the crow flies, to a prairie farm near Nuyaka. For two days at the end of March, 1927, an enormous team of gigantic draft horses pulled the house slowly down the road, outriders preceding and following because it took up the whole width of the road. They could not travel as the crow flies. Instead, their route was about six miles long, and the horses frequently had to stop and rest. At night the house was pulled to one side of the road, and we climbed into our beds, burrowing under the covers to escape the cold because the chimney had been taken down for the move.

I had loved that Rocky Hill farm, and at first I thought I could never live on the bald prairie. Eventually, however, I did learn to love the prairie. I found that one can learn to be happy anywhere. That prairie is no longer devoid of trees. Indeed, parts of it are like a forest today because Marvin planted a thousand trees as a 4-H project.

The little town where we went to high school on the prairie was called Nuyaka. When I mention that name, people always ask what it means in Cherokee. Or Creek. In fact, it doesn't mean anything in any Indian language. In the 1800s, two Presbyterian missionaries came from New York to establish a school for the Creek children. Their students tried to call the missionaries "New Yorkers," but they could not pronounce the name. Instead, they called the two women "Nuyakas," and the school became known as Nuyaka Indian Mission. When the little town was settled nearby, it took the same name.

On the prairie, we worshipped at the little Nuyaka Methodist Church. Our minister taught us to fast and pray like the apostles and always to keep the faith. At Epworth League youth services, we learned about social responsibility: Jane Addams, who founded Hull House in Chicago, and others who lived their beliefs were held up to us as models to be emulated. Before I finished high school, most of our congregation had followed the dust bowl exodus to California, and when the building was sold to the Assembly of God, I continued to worship there. It was still my church home.

Western Oklahoma, the Texas Panhandle, and parts of several other states suffered from lack of rain in many of the years of the 1920s and 1930s, resulting in a series of droughts. The farmers' seed didn't sprout. Without ground cover and rain, the soil became so dry that the winds picked it up, causing dust storms. Much of the Great Depression coincided with the years of the dust bowl, the years John Steinbeck portrayed in *The Grapes of Wrath*.

We lived without electricity, running water, or indoor bathrooms. Our house was lighted by kerosene lamps and heated by wood stoves, and we carried three-gallon milk pails full of water from the spring at Rocky Hill and, later, from the well at Nuyaka. We saved tallow and ashes to make lye soap, washed our clothes on a scrub board and hung them out to dry in the sun. We butchered hogs, salted the pork heavily and hung it in the smokehouse for curing and storage. We canned and dried the fruit from our orchards and hedgerows and stored potatoes in the crawl space under the house that we used as a root cellar. We milked our cows, churned the cream that rose to the top to make butter, and made our own cottage cheese. Our chickens supplied us with fresh meat and with the eggs that we stored with the dairy products in our always-cool storm cellar—because we had no refrigeration.

When we needed cornmeal, we shelled and cleaned corn from the granary for our brother Joe to take to a neighbor's stone mill to be ground—until Joe bought the mill and started grinding corn for our neighbors. We used "everlasting yeast" to make "light bread" from wheat flour. Every time we made bread, we saved a portion of the yeast and carefully fed it with sugar and flour until it was time to start the next batch of bread. We made all our own clothes, except coats and overalls, on the old Singer treadle sewing machine that was run by foot power instead of electricity.

In summer, our house was cooled by the breezes from the open windows. In winter, the heat from the old wood heater and cookstove didn't reach to the outer corners of the room. Bedrooms were always cold; sometimes we would heat rocks or bricks in the oven, wrap them in newspaper, and place them under the covers to warm our feet. Our mattresses were filled with clean, carefully selected corn shucks and topped with soft, warm feather beds filled with the down we plucked from our geese each spring.

Usually there was a quilt in the frame suspended from the ceiling on pulleys. When we needed the space beneath the quilt, we simply raised the frame to the ceiling. Grandmother Perry, who was known far and wide for her beautiful quilting, lived with us until I was six years old. I would sit beside her as she worked, making doll clothes from scraps. If my stitches were too large, Grandmother would have me take them out and redo them until I was able, like her, to make the tiny stitches that use only the bare tip of the needle.

When she tired of sitting or she needed wild herbs or special tree barks for her healing potions, Grandmother would take us for walks in the woods. We waded through clear, shallow streams, and investigated the interior of a huge tree with a hollow trunk big enough for a playhouse. Grandmother told us the names of the birds and animals we glimpsed, as they went about their business. She introduced us to the trees and plants that surrounded us, telling us about the special qualities of each as she explained how she used them in her potions and how she identified and prepared the edible ones. Through Grandmother Perry's eyes, we saw the earth as a bountiful mother who deserved our respect and care.

Grandmother Queen Elizabeth Hutson Perry's family were of English descent and her unusual name and the name of her brother, King Samuel Hutson, reflected family stories of the Old Country. Grandmother spoke an English dialect, and some of it we absorbed unconsciously. On her tongue, "egg" was "aig," and "chair" was "cheer." When Clara and I were juniors in high school, our speech teacher brought this to our attention. For the next two years we worked hard helping each other learn a more standard pronunciation.

Once a year, our family went to town. Much of the year, we all went barefoot, but in early November after the crops were harvested, we would all pile into the old farm wagon for the thirteen-mile trip to Okmulgee to buy shoes. There were no fast food restaurants then. Instead, there was the hot tamale vendor. As I write this, I can see him in my mind's eye, pushing his little vending cart up and down the street, his stentorian voice hawking "Hot tamales, fresh and fine, six for a nickel, twelve for a dime. Somebody's eating them all the time!" My fingers tingle with the memory of that long-ago cold November day, and my mouth waters again at the savory smell and matchless taste of that once-a-year treat.

We all worked diligently to make our living. Mama and my older brother Joe walked miles up and down the fields, one guiding the mule-drawn plow to till the soil and lay out the rows, the other behind another mule, guiding a seeder along the furrow to plant Indian and kaffir corn and cotton. I had never even heard of a tractor and thought of our mules, Rabbit and Dinner Basket, our plow, and our seeder as up-to-date equipment. By the time my red-haired twin sister, Clara, and I were six, we, too, had become farm workers, chopping weeds from the rows of cotton and corn, picking cotton, and harvesting and shucking the corn.

Often exhausted and in pain from long hours of hard labor, we might have despaired, but instead we felt a sense of pride and accomplishment in jobs well done. We revived our spirits and took solace from the simple amenities of life—a drink of clear, cool water, pumped fresh from our deep well; simple, nourishing food; healing sleep; and our love for each other—as we renewed and readied ourselves to tackle the challenges of another day. Through the many hardships and privations of those dust bowl and depression years, we endured and found life good.

During the three years between our father's death and our mother's remarriage, the children of my divorced aunts, Alice Campbell and Mary Capps, whose work took them away from home, joined my sister and me and our two brothers. Margie Campbell and Lillie Capps were the same age as my sister and I, and Clymer Campbell and Everett Capps were two years older. Aunt Alice also left an older son, Kip, with us. Aunt Meda, who lived only two miles away, often brought her daughter, Grace Wheeler, to visit, as well.

I still have happy memories of the five of us little girls playing together, sometimes joined by our brothers. Clara and I were inconsolable when our mother remarried and the cousins went home to live with their mothers. I've often wondered how our mother and grandmother coped with the nine of us, but cope they did. Mama sang as she worked about the house and told us stories of the dancing days of her youth.

One of the dearest treasures in my memory trove is from those years. On a beautiful moonlit night when the sky seemed as bright as day, Mama woke all nine of us and took us out into the yard. In our nightgowns, we formed a circle, Mama in the center. We hopped, skipped and frolicked, while Mama danced lightheartedly and sang a yearning song of her youth, "Oh, the moon shines tonight on pretty Red Wing …"

Not all our days or nights were bright, however. During the days of the dust bowl, there were times when the dust would billow in from the west like clouds. It was said that it even blew across the ocean and settled on the fields and roofs of France and Germany. The day before Clara and I were to go on our high school senior class trip, the dust was so thick one could see only a few feet, and we had to tie wet cloths around our faces to breathe. My mother refused to let us go on the trip.

That night, Clara and I set our alarm clock to go off every hour, and every time it went off we prayed for the dust storm to cease. At five o'clock that morning, dawn broke clear and beautiful with no hint of dust in the air. Reluctantly, Mama told us we could go with our ten classmates. That day we reveled in our first trip to Oklahoma City, our first visit to the zoo, and (in the John Brown department store) our first ride on an escalator.

The history of my family is a mirror of the times. Twice widowed, my mother eventually had six living children. Chester Dugger, her first husband and the father of my older brother, Joe, died in 1911. A rancher, he had been riding on the range and stopped at a country store for a can of meat. The meat was contaminated, and he died of ptomaine poisoning.

Our little brother was named Marvin Beauregard Barefoot in our dead father's honor. (People hearing my maiden name for the first time often assume that it is Bearfoot, a name from one of the northern American Indian tribes. Barefoot, however, is an Irish clan name, and our grandfather Barefoot, tall and redheaded, looked very Irish. However, like many Oklahomans, we do have Native American ancestry. Grandmother Barefoot's father's mother's father was a full blood Cherokee from North Carolina.)

Our mother's third marriage, to Lawrence Gaskill, gave Clara and me two real, live dolls, our little sisters Florence and Paula Maud.

Although we had three different fathers and there were sixteen years between the oldest and youngest, the six of us were very close. Years and distance have not separated us in spirit, even though we have become scattered from coast to coast. Our older brother, Joe, died in 1987, but he will always live on in our wonderful memories of him.

The hardships of our lives pale in comparison to our grandmother's stories of her Civil War childhood in Tennessee from 1861 to 1865. Grandmother told me of her father going to fight in the Confederate Army, leaving Great-grandmother Hutson with their son and three young daughters alone at home. The son was too young to go to war, but because the Union soldiers sometimes killed boys in the South, great-grandmother hid her son in a cave during much of the war.

At one point a group of Northern soldiers descended upon their property and removed the cured meat from the smokehouse and the corn from the granary. They took all the animals—horses, mules, cows, pigs, chickens and even the peacock—before they burned all the outbuildings. As they left, having taken the quilts and blankets and all the food my great-grandmother had, one of the men prepared to set fire to the house.

"Put out that torch," the commander ordered. "That old house isn't worth burning."

Great-grandmother Hutson and her three daughters knew that, despite appearances, the commander must have been a man of conscience, and they were filled with gratitude to be left unharmed and with shelter and the clothes on their backs. After many months, my great-grandfather returned from the war and his son no longer had to seek security in the cave. There was great rejoicing: They had all survived the war and were once more together!

Our other grandmother, Nancy Clara Withrow Barefoot, was very young when the Civil War began. While her father and his seven brothers fought with the Southern army, her mother tended their water mill on the farm that is now Withrow Springs State Park near Huntsville, Arkansas. While once marauding soldiers took all their cured meat, they were generally no threat. However, in the absence of the eight men of the family, my great-grandmother was responsible for the work of nine, keeping the farm work going and grinding corn and feed for their neighbors. Great-grandmother Withrow, six feet tall, broad shouldered and very strong, was ready to rise to the challenge. She would lift sixty-pound bags of grain just as easily as any man. In Grandmother Barefoot's family, the joy of the return of her father and uncles was tempered by mourning for her father's two brothers who died in the war.

My ancestors had survived times of great sacrifice; I, too, had survived, and even thrived, during the hardships of my early life. Surely, I told myself, I could conquer the anger that was destroying me. I could eradicate old prejudices and unproductive attitudes from my mind. PSO's people were my neighbors, and, as my mother had taught me, I must love my neighbors as myself. I had to accept the fact that I was fighting an issue, not people. The issue was nuclear power, not betrayal.

Each day I asked God to direct me and to give me the grace of strength, tranquillity, and inner security. As my prayers were answered, I began to see that the hatred in my heart had driven the health from my body and mind. I began to understand that I must grow into a more objective and mature state of mind. So I prayed that I might learn to treat my adversaries with kindness, respect, understanding—and even brotherly love.

As I progressed in this new discipline, PSO people responded in kind, mirroring my actions and attitudes. The ogres of my imagination became courteous people who treated me with respect. What had begun in an effort to regain the strength I needed for the battle against the nuclear threat, ended as a positive good. By the time hearings convened, I had friends at PSO—friends who, to be sure, were fighting me as fiercely as I was fighting them, but no less friends for all that. During the hearings, one or another of them would often sit down beside me to visit for a few minutes.

Once, during a recess in one of the hearings, PSO President R.O. Newman sat down to laugh about something that had happened in the hearing that morning. At the next recess, one of the members of my group accosted me, "Why are you treating Mr. Newman like a friend. Don't you realize he's the enemy?"

"No," I told her. "Nuclear power is my enemy. If we expect to change PSO's thinking, we must break down barriers, not build them. Mr. Newman may be wrong, but he is my brother, and I treat him with the respect and kindness I owe to a brother."

Later, PSO Vice-President of Finance William O. Stratton, who is about the age of my son, held his hand out to me. "I'm glad to have an opportunity to meet you," he said. "I've always been a great admirer of yours." Then he added, "But you probably aren't interested in what the opposition thinks of you."

Smiling, I shook his hand. "Thank you for your kind words. I *am* interested in what you think," I told him. From that day we have remained friends.

Enlisting Robert

My husband worked so hard at operating the farm and managing the nursing home, that he had no emotional energy left for worrying about nuclear power. I understood how he felt, but nevertheless, when I came across a key sentence or a pertinent paragraph, I would say, "Robert, listen to this," and read it to him. I knew he had such a prodigious memory that anything he read or heard he could quote almost verbatim, and so gradually, against his will, I informed him about the dangers of nuclear power.

The world is a far different place from what it was in 1973, and that difference was partly the cause of a general push to build nuclear power plants—and Black Fox, in particular—and partly its effect. In 1970, gasoline could be bought for less than 35¢ a gallon in the United States and often less than 20¢ a

The I

How many tim
energy policy? Pro
political figures pro
Well, it is importan
heeded, by the volume
First, our waste of all
mined energy efficiency p
and electricity.
Numerous studies, both go
the size of its economy withou
recent conference of engineerin

Material gathered by a National
energy use during the year 2010 whic
by the utility companies.
The Oak Ridge National Laboratory
efficiencies are too little publicized, ha
natural gas consumption by almost 50 p
In a January, 1977, report the Laborator
electrical energy requirements of a typica
nearly 65 percent", with "little or no effect on t
occupants."

or building modifications or equip
ich resulted in savings of almost
the results were "applicable to a la
her structures" used on a 40-hour a
ply of oil, gas and coal in this countr
It is not a question of supply but a qu
ue political influence.
il and gas are three to six times what
s good profits. But the energy monop
roducers' strike by withholding supp
d of at home.

OBSERVER, FEBRUARY 25, 1977

OMA CITY TIMES

risis Spurs Nations

p Nuclear Plants

world's energy pro-
and 83 per cent of
ld's electricity, ac-
to new projections
International Atom-
y Agency here.
apid growth of nu-
wer, if allowed to
without adequate
ional safeguards,

could produce "acute"
safety problems, the agency warned.
In a report on a future
worldwide role of nuclear
power, the agency said the
demand for a large num-
ber of atomic plants might
lead to the manufacture of
components of poor quality

— mass-produced rather
than custom made, as they
are now in the United
States, Great Britain, and
West Germany.
The report, prepared by
four scientists, warned
that national regulatory
agencies, particularly in

gallon in Oklahoma. Most American automobiles got no more than twelve miles to the gallon. Even then, there were people farsighted enough to say that was insanity, but Detroit automobile manufacturers and plastic factories everywhere were convinced that oil would last forever—or at least as long as they did.

Much of the oil to make that gasoline and that plastic was imported from Saudi Arabia and other Islamic countries—countries that wanted to wipe Israel off the face of the earth. They knew that was not going to happen, but at least, they thought, they could get the Israelis to withdraw from the territories they had occupied since the 1967 war. Their weapon of choice was energy blackmail. In October, 1973, a group of them, the Oil Producing Export Countries (OPEC), imposed an oil embargo on the United States and a number of other countries, reducing their oil production by thirty percent. The embargo was lifted after five months, but OPEC made it clear they expected the U.S. to put pressure on Israel to withdraw from the occupied territories.

One man in all this, Sheik Yamani, was, I think, a true world statesman who felt that only a shock of this magnitude would turn the West from its wasteful habits. And he was partly successful; we now have automobiles that get up to seventy and eighty miles per gallon, and people are, in general, much more conscious of energy wastefulness.

How well I remember the effects of the oil shortage: Gasless Sundays, lower speed limits, higher prices, long lines at service stations, and the truckers' strike. National Guard

Grandson, J.J. Dickerson

Of The Energy Crisis

By Ralph Nader

...eard it said that this nation has no comprehensive ...s often as you have heard Jimmy Carter and other ...you one ...winter to summarize what has been learned, if not ...t hearings, studies and investigations since 1973. ...gy in this country is so massive that a deter- ...be our most immediate "new supply", of fuel ...d private, show that the U.S. could double ...more energy per capita than today. At a ...he projection was even more optimistic. ...ciences group portrayed scenarios of ...v lower than the projected forecasts ...ngs on easily attainable energy ...readily homes could cut their ...its engineers reduced "the ...frame office structure by ...working efficiency of the ...vere involved" in the ...ear. The Laboratory ...of office buildings, ...le. ...s and enough for ...e and profits, of ...973 when the ...int more or ...market or

units were called out in eight states to patrol highways and provide convoy protection for nonstriking drivers. Independent truckers delivered seventy percent of the nation's perishable goods, and the increase in diesel fuel prices from 27¢ to 45¢ a gallon ate into their profits in a serious way. Two truckers were killed, and scores were injured by gunfire, stonings, beatings, and burning rigs. Bomb threats were reported in more than twenty states.

In two years, the price of OPEC oil rose from $1.70 to $10.12 a barrel,[1] depressing the world economy. The United States, reasonably enough, refused to allow itself to continue to be hostage to the political, religious, and economic whims of other nations, and President Nixon and Congress started looking for solutions. They made a commitment to building 1000 nuclear reactors by the year 2000, including a number of breeder reactors to produce weapons-grade plutonium.[2] By January, 1974, 42 nuclear plants had operating licenses, 56 had construction permits, and there were another 101 plants on order.[3]

President Nixon asked Congress for $53 million to fund reactor safety research, but that was a drop in the bucket compared to the $207 million he wanted in research and development funds for the extremely dangerous liquid-metal fast breeder reactor![4]

Not everyone thought Nixon and Congress had their priorities straight: "If we should place most of our energy research-and-development eggs in the nuclear basket we would find ourselves with problem-prone nuclear power as the only alternative to oil and coal as our major source of energy," warned the *New York Times*.[5] Even before the OPEC crisis, the National Science Foundation and National Aeronautics and Space Administration had concluded that solar energy, conservation and improved technology should be pursued at least as vigorously as nuclear energy.[6] They believed solar energy could economically supply thirty-five percent of the nation's heating and cooling energy and twenty percent of electrical power by 2020.

By 1974, however, President Nixon's wishes were increasingly beside the point. On August 9, 1974, the House Judiciary Committee voted to impeach him for his complicity in an illegal raid on the headquarters of the Democratic National Committee, and Nixon resigned. The country went into an orgy of national shame and grief over that black mark on our national integrity.[7]

Before 1974, only three nations—the United States, the Union of Soviet Socialist Republics, and the United Kingdom—possessed nuclear bombs. That year saw the beginning of the proliferation of nuclear weapons that now threatens the whole world. India exploded its first nuclear bomb, and Pakistan began pushing its own nuclear program to counteract the Indian nuclear "threat," while France and China exploded nuclear devices on the same day in June.

This was the background against which the Black Fox drama was to be played out.

By August of 1973, I had three months of research behind me, and I felt I had done my homework. I now knew that the plants themselves were unsafe. I also knew there was no safe way to store the highly radioactive by-products of nuclear power generation. I had learned that it would be relatively easy for terrorist groups to steal enough to contaminate the lands of their prey—or even to make bombs to use against them. Since some of those radioactive by-products last for millennia, we could be helping to poison the whole earth—not only our own children. "I can't keep quiet any longer, Robert," I told him. "It's time for me to start talking to people."

"You can't start a fight against big business and the government," he objected. "Your life could be in danger."

"I'm fifty-six years old," I countered. "I've lived twenty years longer than my father. In many countries, I would be considered ancient. I can't just sit and hold my hands. Besides, you're exaggerating the danger: PSO and our government will deal me no harm."

Silently, we stood side by side at the window, watching our little grandson J.J. as he played in the yard. After a moment, I continued, "J.J. isn't even two. If I can't protect him, then my life is not worth living, anyway. You know that radiation affects the dividing cells of the body most and that younger children have more dividing cells, so the younger the child, the more likely its health will be affected. J.J. could develop leukemia from routine emissions of radioactive elements from a nuclear power plant!"

I told Robert about a child with leukemia I had cared for as a nursing student. The five-year old was suffering and too weak to walk. He spent weeks at a time in the hospital. "There are many causes of leukemia," I said, "but if we can prevent one of them, we will have justified our existence. Radiation poisoning can even cause birth defects. Would you be able to live with yourself, if, through inaction, we were responsible for allowing a future grandchild to be born with such an affliction?"

I had sounded the right note: The fond grandfather's attitude changed. "But you can't just take nuclear power away," he cautioned. "You've seen all the articles about an energy crisis. Even if the whole thing has been contrived to raise the price of oil or to make nuclear power more attractive, we have to offer a viable alternative if we are to have any hope of success. People are too frightened about the future."

Suddenly, I was excited. "Do you remember that Jacobsen Windcharger Lawrence installed to light Mama's house when we were in college?" I asked him.

Robert, too, caught fire. "Why don't we go to Stillwater and find out if there's any current research on wind power going on? Maybe even solar power, too!"

An Honest Man

Robert and I always loved to return to Stillwater. There, we had worked and studied at Oklahoma Agricultural and Mechanical College. There we had spent most of the years of our courtship and early marriage. There we had returned periodically when our children were in 4-H Club.

As teenagers Robert and I were both active 4-H members. The summer I graduated from high school in 1935, I won first place in the Girls' Timely Topic Contest at the State 4-H Club Round-Up at OAMC. I later discovered that Robert, too, had been a delegate. That fall, he and I were among forty-nine Oklahoma 4-H members awarded a trip to the National Farmer's Union Convention at Kankakee, Illinois. We had been winners of the Northeast Oklahoma District Fair 4-H speech contests, Robert at the Tulsa State Fair and I at the Muskogee State Fair.

Most of the 4-H members were already aboard the train when I boarded at the siding in

Carrie & Robert, November, 1937

Oklahoma 4-H Club

TIMELY TOPICS

Oklahoma Four-H Club

FUTURE HOMEMAKERS OF OUR NATION

CARRIE BAREFOOT, Okmulgee County

Sooner or later most girls marry. Statistics prove that ninety percent e women of our nation at one time or another, in the course of their assume the responsibilities of a home. Since time immemorial mar- as been a partnership business, the accepted custom being that labor by the sweat of his brow to provide the concrete me Clearly, if he does this and does it well, he has the inner workings of the household. The privilege of taking a house and maki h the ages, a nation has be e homes remained wh g unit, the nati ne ate an

Morris. When we pulled into the Claremore station, a handsome young man was waving good-bye to his proud parents. In that moment, I knew that I had seen my future husband!

I began to wonder about my "future husband" when I saw him sit down to play cards with some other member of the group. My mother had always insisted that card playing was the work of the devil, yet I allowed myself to be drawn into a game. Robert was almost as shy as I, but he taught me to play well enough that with beginner's luck and his help, I won a game. Then guilt and remorse overcame me. That night on my knees, I pled for God to forgive my sin.

As we entered Illinois, the train stopped for a few minutes, and all the boys filed out of the train and into a saloon across the street, the first saloon I'd ever seen. Each of them—including Robert—came back with a small bottle of whisky prominently displayed in his hip pocket. I thought to myself, *that* can't be your future husband!

Before Oklahoma could become a state in 1907, a new constitution had to be written. Because alcohol had allowed whites to rob them of the land that had originally been theirs, the Native American tribes, whose land then comprised about half of Oklahoma, refused to give up their sovereignty unless alcohol was made constitutionally illegal. Thus, it was unlawful to buy or sell alcoholic beverages in our state. Prohibition did not still the craving, however, and whisky was made and sold illegally. "Revenooers" often found barrels of fermenting "mash" and whisky stills hidden in the back woods. I knew families who were destitute because fathers spent their money buying "moonshine" whisky. Sometimes the moonshine was diluted with wood alcohol, which caused a debilitating nerve disorder, called "Jake Leg," that made it impossible for its victims to walk normally. When the moonshiners could be found, they were tried and prosecuted, but nothing could repair the damage they had caused.

I sat there contemplating those horrors and refusing to meet Robert's eyes. Finally, he came over to me and asked why I was so quiet. Bolstering my resolve, I reminded myself of the words of George Washington that Mama often used to admonish us: "Those who value their reputations should choose carefully the people with whom they associate." With all the righteous indignation of a Carry Nation, I answered quietly, but emphatically, "I don't associate with people who drink."

Robert laughed. "I have never yet taken a drink," he said, "and I never intend to. That bottle is only a souvenir for Mom's medicine cabinet." Years later, a small swallow from that same bottle of whisky saved the life of our little daughter Florence when, before the days of antibiotics, she was suffocating with croup. That near tragedy taught me the value of questioning preconceptions and dogmas.

We all spent a day sight-seeing in Chicago before going on to the convention in Kankakee. As I got off the train, Robert took my suitcase and carried it along with his. I hoped that this was a sign that maybe he had seen something in me that he liked—that perhaps the attraction was mutual! A plain girl with my long, dark hair braided and wrapped around my head, I wore no make-up, and I had never flirted or dated. Pride forbade me to let him see that I was attracted to him. I treated him as I would my brother.

I wouldn't have felt so flattered if I had known then what Robert's father was to tell me almost thirty-five years later: In preparation for the trip, he had told Robert that since the two of us were the winners in the same category in our district, it was his gentlemanly duty to carry my suitcase and look after me. He was only following his father's instructions!

At the convention and on the trip home, Robert and I were often together, but when we parted there were no promises of letters or future communication. Undaunted, I knew what I had to do: I would get my degree and go to Claremore to teach school. I was going to marry Robert Dickerson!

In the spring of 1936, I worked for board and room in the home of an Okmulgee family so I could attend Okmulgee Junior College, and my brother Joe sold a pig to pay for my books and expenses. By then, Clara had already been at A&M for a semester, working for her board and room in the home of a judge and his wife, and that summer she accompanied them to their summer home in Colorado Springs.

While she was in Colorado Springs, I was again at the State 4-H Round-Up in Stillwater. Walking across campus with the other

Okmulgee 4-Hers, I recognized Robert's back as he walked in front of me. My surprise at seeing him drove his name completely out of my mind. I didn't want to encounter him without greeting him by name, so, embarrassed, I turned to walk the other way. Moments later, I felt a tap on my shoulder and a voice saying, "Hello, Carrie." He hadn't forgotten my name!

At the campus drug store, we did a lot of catching up as Robert treated me to my first banana split. When he told me he planned to attend Oklahoma A&M College in Stillwater in the fall, I immediately decided to advertise in the Stillwater newspaper for a board and room job so I, too, could attend A&M.

We were still in the depression and dust bowl days, and my family had no money for college, not even for transportation to college. Fortunately, tuition was free, but I wondered how I would ever get to Stillwater to begin the room-and-board job I had found. One day, the society column of the Okmulgee newspaper mentioned that another Okmulgee Junior College student was planning to attend A&M in the fall and that his parents would drive with him to Stillwater. Swallowing my pride, I walked a mile to the home of our only neighbor with a telephone, and asked my classmate's mother for a ride. That was not the first or the last time I have had to swallow my pride, but it has never gotten any easier. With everything else arranged, my brother Joe sold another pig to get the money for my books and expenses.

Robert and I worked hard at our jobs and at our studies. He made straight A's. In spite of being out of school more than I was in, picking cotton and doing other farm work, I had been a straight-A student in high school and valedictorian of my high school graduating class, so I was disappointed to earn B's with only an occasional A in college.

I had agreed to work three or four hours daily for my room and board, but the responsibilities I was often given required eight hours—or much more—each day. One of my duties was to bathe the family bird dogs in the bathtub. I was dumbfounded; our dogs had always gone for a swim in the creek when they felt the need. Another chore I had never before done was to butcher and prepare a turkey for Thanksgiving dinner. I had been butchering and preparing chickens since my early teens, but after that turkey experience I decided it would be better to be a vegetarian! Caring for a new baby, however, was something I had had lots of experience with, and I enjoyed every minute with their precious little boy.

Most nights, I went to bed with my chemistry book tucked under my pillow. Unfortunately, none of the information in the book seeped into my brain while I slept! I promised myself that someday I would go back to school and take chemistry over again just for fun, but thirty years later when I decided to become a registered nurse, I was informed that because I had sixteen hours of chemistry, I did not need to repeat it. I enrolled in the course anyway because I wanted to understand the new ideas that had emerged from the intervening thirty years of research. No other "A" has ever made me so proud!

Few students had cars in our college days. Wherever Robert and I went, day or night, in snow, sleet, rain, or sun, we walked. Our dating consisted of attending the Methodist Church worship services on Sunday mornings and evenings. Occasionally we visited my twin sister or, when we had a little extra money, went to a movie.

Robert worked in the office of state 4-H Club director, B.A. Pratt. Once he took on an extra job, digging ditches on the campus. When he received his paycheck, he phoned to tell me that he had been paid more than he figured he had earned. Many people would have taken the extra money and said nothing; Robert checked and confirmed that there had been a mistake. The next day's college newspaper headline told Diogenes to put away his lantern because they had found the honest man! There are no words to describe the pride and love I held for my future husband!

Robert and I were married before we finished college, but we both went on to earn our degrees, his in animal husbandry and mine in home economics education. In 1938, we moved to the beautiful Claremore farm where he had been reared as an only child after his little sister Paula died at the age of two, and there we brought up our son and three daughters. I have lived happily on that farm all these years.

Robert's father, Eugene had come to Claremore from northern Missouri in 1908, when the city had hired him as its school principal. In 1909, he had become Claremore's first superintendent of schools and instituted a much-needed building program, and in 1910, with the help of a long-term government loan, he bought our farm. He married school teacher Anna Hanes, a member of a prominent Sageeyah Cherokee family, and they continued to teach for a number of years before moving permanently to the land where he and Robert eventually established an organic farm and a grade-A dairy.

For many years, I, too, taught elementary school. After spending several of my summers working for a master's degree, I taught secondary school home economics in Rogers and Mayes County schools. Finally, my increasing knowledge of the benefits of healthful foods, grown without pesticides, prompted me to open the organic bakery and natural food store I operated for eight years and to study dietetics.

Many of my customers had frail, elderly parents who did not eat properly—or were themselves frail and elderly. Robert and I came to recognize the great need for healthful, loving care for those who had reared our generation and were no longer able to care for themselves. We became determined to build a nursing home that would be a place of light, comfort and cheer and where palates would be tempted with the savory foods of their youth. Although I was never able to spare the time to complete the year's hospital internship that would have qualified me for membership in the American Dietetics Association, the Oklahoma Department of Health granted me a special certificate that allowed me to act as dietitian for our nursing home.

Like most nursing homes, we employed LPNs, because RNs were rarely available. Then in March, 1967, at a state nursing home convention in Tulsa, a bombshell was dropped on us: We would be given three years grace before we had to have RNs as nursing supervisors. The same thought was reflected on nearly every face in the room: *Where are we going to find them?* Then I realized that three years is the time required to qualify as a registered nurse. I pushed back my chair from the luncheon table and went out into the hall to telephone Hillcrest Hospital.

"We don't accept students over the age of fifty," I was told. I had two months to spare! Along with our daughter Florence and niece, Martha Jane Cahalen, I enrolled in Hillcrest Medical Center School of Nursing in Tulsa, and in 1970 the three of us became full-fledged RNs.

If Windmills Were Gold-Plated

All "renewable" sources of energy—wind and water, burnable organic materials (like wood and buffalo chips), and sunlight—and fossil sources—oil, gas, and coal—are gifts of the sun, compatible with human life. Nuclear power, however, calls upon energies too fierce to be compatible with stable life forms. While there are beneficial mutations, the vast majority, like cancer, are lethal. Sunlight and petroleum products can produce lethal mutations, but it is possible to avoid both of them; theoretically nuclear radiation, too, can be avoided but few of us can afford to live in lead boxes—even if we wished to do so.

The Daily O'Collegian Tuesday, February 8, 1977 Page 6

Turbine produces ele

Wind en

By LIZ KEYS
Staff Writer

"...and with the wind chill factor the temperature is dropping to 100 degrees below zero."

Our neighbors to...
experiencin...

a 30-foot, bicycle whee...
turbine.
The t...

TW-4-3-77

Solar Energy Can Sup...
Major Needs, Study Say

...the sun's clean ...supply 40...

But so far it has been generally considered a distant and probably minor source of energy.

Hayes challenges that view. Hayes argues that solar power may never because it requires costly ...claims it could meet ...'s energy demands are...

processes and other energy ...use the kind of energy off... sun: widespread, low-inten... available only during ... sharply reduced by cloud...

But Hayes says large ...ergy even in industry ...sun could provide ...low temperature hea... ing point of water.

for a h...

has
e years
hines.

Robert and I wanted to know how viable non-fossil, non-nuclear sources of energy were. For a century, Oklahoma citizens have been going to Stillwater to find out what they needed to know, and the faculty at the university there have been doing their best to respond. By 1973, Oklahoma A&M had long since been renamed Oklahoma State University. The campus looked very different from the way we had first seen it, but the faculty came through for us just as it always had.

On this trip, Robert and I consulted Dr. Jack Allison at OSU's electrical engineering department. We found that long before the "energy crisis" the department's researchers had been developing energy systems involving wind, sun, and hydrogen power, separately and in combination—both for individual residences and for cities. We also found that PSO had funded $10,000 worth of the research!

Why, we wondered, would they want to build an expensive nuclear plant, when they could harness the wind much more cheaply? Was it possible that all the technology we were seeing was merely window dressing and that PSO was right in saying that these sources of energy were impractical?

Several months later, at a safe energy seminar at the University of Oklahoma at Norman we discovered the answer. "In most cases," we were told, "the utility company grants given for research in solar and wind energy have the express purpose of proving these sources impractical."

Because utility commissions set rates to provide for a specified rate of return on investments, the greater the investment, the greater the return. If a nuclear power plant costs three times as much to build as wind generators producing the same amount of electricity, they naturally pick the more expensive technology. "Maybe if windmills were gold-plated," the speaker concluded wryly, "the utilities might be interested in wind power."

That day at OSU though, we were thoroughly impressed by the ingenious ways people like Drs. Allison, Roger J. Schoeppel, Richard Murray, and William Hughes had been solving technical problems. The windmill itself is mechanically simple, efficient, and relatively inexpensive, but it usually doesn't supply the constant frequency and voltage that modern electrical devices require. OSU's electrical engineering department had produced a reliable generator that supplied constant frequency and voltage no matter how fast or how slowly the windmill turned.

A more daring direction for their work was the use of hydrogen to store energy and then release it safely. Hydrogen, the ninth most abundant atom on earth, releases 97% of its stored energy when burned, and because its burning produces only water, it releases no chemical pollutants and does not contribute to greenhouse effect. Chemists and engineers have been aware of the advantages of this fuel for decades. Why then, we asked, is it not more widely used as a fuel?

We found that there are two principal reasons—expense, because it is produced by the hydrolysis of water (which requires electricity) and safety, because under normal circumstances it is explosive. We also learned that OSU engineers had solved both problems. Because it is such an efficient way to store energy, hydrogen can be made with excess electricity at times of low demand, saving 97% of that energy for later use; that solves the problem of expense. The problem of safety is somewhat different.

One day when I was about nine years old, I stayed home from school to pick cotton with a group of neighbors. As I reached for the next boll, I heard a cry and looked up to see a neighbor throw up her arms and shriek, "The world is coming to an end!" Following her gaze, we saw an unearthly sight. An enormous, elongated object seemed to fill the heavens. It appeared to hover motionless where the trees on the hill met the sky and then very slowly drifted out of sight.

While our neighbor stood there waiting for the heavenly trump, my sister Clara was delightedly watching the same sight from the tall, wide windows of the Rocky Hill School. I began explaining to our neighbor what we had read in the newspaper: "It's a zeppelin!"

In the 1920s, Graf Zeppelin developed a number of dirigible lighter-than-air craft borne aloft by enormous, elongated helium-filled balloons. They were originally safer and more efficient than airplanes because of the great lifting power of the inert gas that filled them and because they would not fall out of the sky if their engines stopped. When war broke out in Europe, however, helium became too expensive; it was replaced with the highly flammable gas, hydrogen. In 1937, the zeppelin Hindenburg ignited and burned just before landing. Scores of people burned to death in full view of loved ones and reporters, and hydrogen was firmly established as an untamable monster in most people's minds.

On May 6, 1962, the twenty-fifth anniversary of the Hindenburg disaster, Dr. Schoeppel and ten other scientists and businessmen founded the Hindenburg Society, dedicated to "the safe utilization of hydrogen as a fuel." Modern technology had made the problem of hydrogen's explosiveness less overwhelming. With a 1968 grant for hydrogen research, OSU professors had successfully converted several small one-cylinder engines and one large experimental prototype to run smoothly on hydrogen.

Automobile manufacturers and oil producers, however, have a vested interest in gasoline engines, and through their lobbying efforts, funding sources for hydrogen research had dried up.

The spring after our visit, we invited Dr. Allison to speak at a Sierra Club meeting in Tulsa. About 200 people attended—including several PSO dignitaries who sat in the front row with their tape recorders. "If PSO wished," Dr. Allison said, "they could make the same amount of electricity at Inola at one-third the cost, simply by installing windmills instead of the nuclear generators."

Council Reviews
Plant Oppositio

CLAREMORE (Okla.) Pr

Claremore city officials are studying the possibility of opposing the location of a nuclear power generating plant near Inola.

After a presentation by Mrs. Carrie Dickerson, owner of Aunt Carries Nursing Home in Claremore, the council voted to u

"God made rattlesnakes, too."

Assured that there were viable sources of renewable energy available, Robert became my most ardent supporter, and I felt free to start campaigning. I made copies of pertinent information and handed them to everyone I met. And I talked to everyone I met—in the nursing home, on the street, in the grocery store, the post office, and the beauty shop. Then I started looking for people and organizations who could help me get the message out.

One of my father-in-law's elementary school students was Bill Briscoe. Bill grew up and went to fight in World War II. When he came home, he became one of Robert's students in a federally sponsored agriculture class for returning veterans. A long-time Farmers' Union member, he served on their state board with Robert's father and eventually became our state representative. We knew that if he supported us, he could make a great deal of difference, so Robert and I went to his office to talk to him. When he heard what Robert had to say, Bill, too, became concerned about the plant. Throughout the years to come, Bill and his wife Ruth remained a wonderful source of moral and practical support and advice.

ola

Proposed PSC Plant Si

THE

TOBER 4, 1973

B ____PH MARSH
____l Bureau
Tri ____ Resi-
OK
den ____
wo ____
ti ____

plant were put in, he probably would absorb only an added 1 to 10 units in addition to natural radiation for the rest of his ____ Smith said.

____ WAS testifying for the ____ervice Co. before the ____ma Corporation Com-n. The firm has filed an ____cation for approval of a nuclear near Inola for a nuclear ____erating plant for electrici

lery o
"W
activ
viro
p
c

"Radiation emitted from ____ nuclear power plant is ____ tremely low," Smith told

SS WEDNESDAY SEPTEMBER 5, 1973—Page 3

nted" at an upcoming ____ to have the authority to ____ the city attorney to be

oers of the committee ____ Albert Hardison, Bob and H.P. Keeling. oroposal passed by a ____us vote.

oppose location of the ____ within 25 miles of Clarer ____ A pre-hearing confer ____ before the Oklahoma ____ poration Commission has ____ set for Monday at 10 a. ____ Oklahoma City. The full-f ____ hearing will be at the sam ____ and place Oc___ 2

My next stop was at the *Claremore Progress*. Editor Donn Dodd listened courteously but, I felt, with a closed mind. Just how strongly he opposed my stand soon became obvious in the editorials he wrote regularly over the years, attacking my actions and words. When Black Fox was finally canceled nine years later, his editorial predicted, "Future generations will say Carrie Dickerson caused them to freeze in the dark!"

I went to the headquarters of any organization I came across, trying to get their support and cooperation in the battle against Black Fox. It wasn't easy to do; I would often be given a tongue-lashing before I had even had an opportunity to explain my stand, so it was with trepidation that I knocked on the door of the Tulsa Indian Emphasis Program, one of several offices in an old brick school building at Sixth and Peoria in Tulsa. I feared the door would once again slam in my face. Instead, a beautiful Pawnee wearing a long, thick black braid, June Echohawk, opened the door and welcomed me with open arms. Sitting at a table across the room was a young Choctaw at work on *The Tulsa Indian News*. That was the beginning of a long, warm friendship with two wonderful people. June and I still work together on various projects, and Dan Agent, the young Choctaw, went on to establish the *Tulsa Free Press*, which for years included an article in just about every issue explaining why Black Fox should not be built; he is now at the Smithsonian Institution in Washington.

I joined several clubs and organizations, hoping to persuade their members to come out as a group in opposition to nuclear power, but PSO people were already influential in most organizations—even in church groups. It had not occurred to me that I could organize my own group.

At about the same time Robert and I were on our way to Stillwater, PSO sent a formal request for approval of the proposed plant to each city council and city commission within a twenty-five mile radius of Inola. The Claremore City Council scheduled a vote on PSO's request for September 4. Fearing that if the city leaders approved the plant, the citizens would follow suit, I visited Mayor Jack Marshall early that day and was given permission to make a statement to the council before their vote. After I spoke, I presented each of the council members with a packet of information on nuclear power.

My talk at the council meeting was my first public appearance to talk about nuclear power. It was also the first time I had ever attended a city council meeting. I reminded the council members that, "As businessmen, you don't make a decision without studying the facts," and then went on to suggest that, "you table the motion to approve the proposed nuclear facility until you have done the research on the subject." I also suggested that they send a representative to the Oklahoma Corporation Commission's October hearing on the same PSO request. One of the members made a formal motion that the city attorney be that representative, and the proposal passed by unanimous vote.

To my astonishment, "Council Reviews Inola Plant Opposition," was the front page headline in the next evening's *Claremore Progress*.[8] The article reviewed my reasons for opposing the plant, including its effects on public health (increased incidence of leukemia, cancer and birth defects), the pollution of air, water and soil, and the greatest problem of all, the disposal of radioactive wastes.

It then quoted PSO public relations director Joe Bevis saying, "She has the right to her beliefs, but we certainly can't agree with them. We believe firmly this type of power is safe and clean. All her claims have been disproved by both the AEC and independent studies." However, Bevis also admitted, "Nuclear waste is a problem. By the time the plant in Inola is opened, it is hoped a real good solution to the waste problem will be available."

I was outraged. PSO was proposing to produce the most hazardous waste known to humanity—some of which would not dissipate for tens and hundreds of thousands of years—and the best they could say was that they "hoped" that a "real good solution" would be available by the time the plant was built! How could the AEC be so irresponsible as to allow PSO to request a building permit when there was no satisfactory solution to be found?

I also felt that the newspaper had sabotaged me. In reality, because I was virtually unknown, the article was probably the best thing that could have happened to me. That newspaper story launched my crusade—and our organization!

The first call came from Hugh Garnett at KWHW in Altus, Oklahoma. He asked me the name of my organization. I was unprepared for the question, but thinking, *If I admit I'm alone, I'll be a dead duck,* I answered without hesitation, "Citizens' Action Group." A few days earlier I had been visiting with one of the residents of our nursing home when on her television I had seen Ralph Nader talking about the accomplishments of his "Citizens' Action Groups," so the name was ready to pop into my mind when I needed it.

I had read a news release in which Nader was quoted as saying, "It is morally outrageous to create a radioactive legacy which will mortgage the future for the next fifty generations in exchange for a little electricity today. Americans should become informed and involved. With thirty nuclear power plants now in operation (and more planned) few people will be more than a few miles from a potential radioactive holocaust. Because nuclear reactors are unsafe and pose unacceptable risks to the public the nuclear power program should be brought to a halt."

I told myself, *Ralph won't mind if I borrow his group's name for a while.*

My mother had always taught me to never tell a lie—that if I did, I'd have to defend it with a second lie, and she was right! Garnett's second question was, "And how many members in your organization?"

I breathed a quick prayer, *Oh God, don't let this be a lie. Please bring forth one hundred people who believe as I do!* Offhandedly, I answered, "Oh, about a hundred."

The news of my opposition to the nuclear power plant and of our newly organized, one-hundred-member group, went across the air waves. Reporters from newspapers and radio and television stations called, asking for more information. People called and wrote, asking to become the one-hundred-and-first member of Citizens' Action Group. The first, Carolyn Henning of Claremore, immediately started helping me with the secretarial work. In no time, my prayer was answered and there were over a hundred members! Robert and I were no longer alone.

One day, three acquaintances came to tell me that as a Christian, I shouldn't be fighting the government, that I was wrong to oppose nuclear power. "God made uranium for us to use," they told me. "When you oppose its use, you are going against God."

"God made rattlesnakes, too," I told them, "but he also gave me the sense not to step on one!" I went on to tell them that ours is a government of the people, by the people, and for the people. "As one small part of that government," I said, "I have a duty to speak up when we are going in the wrong direction."

One of the most influential people in my early life was Okmulgee County Home Demonstration Agent Norine Hughes. A shy and self-conscious child, I was inspired by her belief in me. "Never sit on the fence," she would say. "You'll step on toes and make enemies when you stand for something, and you won't be as popular as the fence sitter. But the friends you do have will be worth having. Stay off the fence and make your life count!"

Without her teaching, I would have found it much harder to deal with the threatening telephone calls I received. Prominent businessmen I had known during my teaching career phoned and called me terrible names. One screamed, "Why don't you put your apron on and get back in the kitchen where you belong?"

At first I couldn't understand why they were behaving so uncharacteristically. Eventually, I realized that they probably owned stock in PSO's parent company, Central and Southwest, and possibly uranium stock, as well. They must have become hysterical when they felt their pocketbooks were being threatened. Each time one of them called, I spoke as kindly as possible and invited the caller to my home to discuss the issue at more length with Robert and me. That would be the last I would hear from them.

Robert's parents, in their late eighties at the time, were living in our nursing home, and his Aunt Ruth often visited my mother-in-law and another sister, Sonora, there. Aunt Ruth was very active and influential in the community and belonged to all the important civic and social organizations. She believed government and big utilities could not be wrong and I was just impeding progress.

I was a great embarrassment to Aunt Ruth. One day she came to the nurse's desk and asked me, "Do you know what the people are saying about you in the meetings I attend? They're saying you're a *radical*." I was too surprised to do anything but laugh. Angrily, she protested, "It's nothing to laugh about. You're destroying our family name!"

Knowing she sincerely thought me wrong, I was able to be patient with Aunt Ruth, and as time went by she came to see that I was right and began to invite me to speak at her club meetings.

Friends and teachers with whom I had taught began to turn and walk away as fast as they could when they saw me coming down the street. I began to feel ostracized. I know that some of them must have resented being forced to listen to what I had to say, and others probably wanted to know who I, a registered nurse, nursing home administrator, and former school teacher, thought I was to challenge the "experts" at PSO. Was I being presumptuous in expecting people to listen to me?

This severe treatment prompted me to examine my own motives very closely and to take care that my own ego didn't get in the way. What did it matter if my feelings were hurt? Building up my ego wasn't the goal. My goal was to stop the nuclear facility from being built. When the criticism cut me to the quick, I reminded myself that I was an instrument, a catalyst, in a greater cause. What did it matter if people thought I was radical or an alarmist or off my rocker? I knew I wasn't.

Environment[ ]
Tickets to Go

Tickets will go on sale Thursday at the Metroppolitan Tulsa Chamber of Commerce for individual events of the National Forum for Growth with Environmental Quality beginning Monday at the assembly Center.

Tickets are $5 each for the panel debates, $10 each for three luncheon events, and

Above: Carrie at a Sierra Club meeting with Ruth Myers & Dan Fitzgerald

Organizing

That September was busy. I rented a booth at the three-day Rogers County Fair, where I handed out information sheets, visited with people and gathered signatures on a national petition against nuclear power. Responses were mixed. One woman shook her fist in my face and said, "I've worked hard to get the money for my appliances, and now if you have your way, I won't have enough electricity to run them!"

One day, I looked up to see a dozen young men from Inola standing shoulder to shoulder in front of my booth. They called me names and told me loudly and offensively that I was interfering with Inola's progress and that I had no business meddling in their affairs. I was too frightened to speak. At that moment, Inola's mayor, my former student Tommy Dyer, appeared. Tommy put his arm around me. "Nuclear power or no nuclear power, I'm still your friend," he assured me.

Tommy listened attentively as I told him about the drawbacks of

LET THE FREE MARKET PAY FOR NUCLEAR POWER (not taxpayers)

Doris Gunn at an Earth Day rally in Oklahoma City, 1981.

TULSA DAILY WORLD, SUNDAY, SEPTEMBER 23, 1973

Attendance May Exceed 600 For Environment Forum Here

An attendance of more than 600 is expected for a national forum beginning Monday at the Assembly Center to seek a middle-of-the-road philosophy on how to have economic growth and still maintain a high level of environmental quality.

Related News On A-22

azine. The Honolulu Advertiser and Star-Bulletin is sending its environmental editor.

McGraw-Hill, the nation's largest magazine publisher, is sending a staff writer from its Washington, D.C. office.

The state's public radio station, KOSU at Stillwater, will tape all panel debates and speeches for rebroadcast, and some of the recordings are expected to be used on national radio.

Washington, and technology, Dr. Philip L. Johnson, director of Environmental Systems and Resources Division, National Science Foundation, Washington.

Special speakers include Jules Bergman, ABC News science editor, Monday luncheon; Russell W. Peterson, former governor of Delaware, Monday banquet; Christian A. Herter Jr., special assistant to the Secretary of State for environmental affairs, Tuesday

nuclear power, but I knew that, as mayor, he had to tread lightly. Many of his constituency were looking forward to jobs building the plant and to sending their children to "the best school in the state," financed by taxes from the nuclear facility. I wanted fine schools, too, but in a safe environment.

Among the fairgoers who joined with me were Deborah Roach and Terrie Neely, as well as Lewis Bean, a retired school principal and state legislator, and his wife, Violet. All became stalwarts in our citizens action group, CASE (Citizens Action for Safe Energy), holding meetings in their homes and helping with signature gathering and other activities. The Beans later provided a place for expert witnesses to stay when they came to testify in the hearings.

That month, there was also a three-day national seminar on energy and the environment in the elegant old Mayo Hotel in downtown Tulsa. I suspect Tulsa was chosen to give PSO a boost, because while it was done subtly, the positive aspects of nuclear power were the focal points of the seminar.

The eighty-five-dollar fee I paid to attend was worth it, because that meeting was the beginning of my long and productive collaboration with Ilene Younghein of Oklahoma City. I had met Ilene two years earlier at a Clean Air meeting in Norman, and when at the seminar luncheon I overheard her talk about going to the Oklahoma Corporation Commission hearing on the plant, I was ecstatic to think that someone else was interested.

"Oh, Ilene, are you really?" I interrupted her conversation.

"What of it?" she demanded truculently. I was taken aback. However, when she realized that I was asking because I was truly excited, she explained that she had thought I was delivering just one more of the rebuffs she had come to expect when she talked about her opposition to nuclear power.

Ilene had been studying nuclear power much longer than I, and when I realized the extent of her knowledge of radiation and its effects, I felt like a novice. In recent months, she had been checking into the Kerr-McGee Cimarron Facility, a plutonium processing plant near Crescent, Oklahoma, twenty miles north of Oklahoma City. She had become convinced that the plant was polluting the environment in the area near where she lived. For the next nine years, as my co-chair of CASE, she provided the brain and I the brawn.

Ilene introduced me to Doris Gunn of Oklahoma City, who represented the League of Women Voters at the seminar. Doris was an enthusiastic proponent of solar energy. Her office was a spare bedroom in her home a block from the state capitol building. Just like my living room and study at home, bulletins and books on renewable energy and nuclear power covered every available space, and the walls were papered with newspaper clippings. Doris became deeply involved in CASE, and we often met at her house when we were lobbying or attending hearings at the capitol.

The seminar was also the scene of my first meeting with Tulsa lawyer Tom Dalton. A brilliant researcher and a creative thinker, Tom was a leader in the Oklahoma chapter of the Sierra Club. He was to become one of the most effective workers in our struggle. There, too, I met another of the Sierra Club leaders, OU geography professor Marvin Baker. Since May, when I had first learned of PSO's plans, I had spoken frequently on the telephone with both of them, but I had never before met either in person.

Tom's telephone voice was gruff, and I would not have had the courage to make a second call had I not been so afraid of nuclear power. As I came to know him, I learned that his voice belied a gentle personality. Tom was as concerned as I about the devastating effects of nuclear power, but we persuaded him only with great difficulty to represent CASE when we first needed him. Once he had become involved, however, his brilliant, subtle, and tenacious mind became a mainstay of our efforts. I firmly believe that without Tom, we would have lost in the very first round of hearings and Black Fox would long since have been in operation.

The October 3, 1973, Corporation Commission hearing turned out to be less than world-shaking, although there were several interesting moments. The most notable of these came when Chairman Charles Nesbitt asked Jack Wells, PSO's vice president of engineering, why PSO had come to the OCC for approval when it was common knowledge that only the AEC had jurisdiction over approval of a nuclear power plant. Mr. Wells' face grew so red and the veins on his neck stood out so that, alarmed, I wished I had my blood pressure cuff handy.

Chairman Nesbitt persisted, "Is PSO seeking approval from the OCC to silence the protesting voices of the citizens like these?" he asked, motioning toward us "dissidents"—including Oklahoma City attorney James Ikard, Sierra Club member Deborah Allen, Ilene and Gaylord Younghein and Robert and me—who had come to testify.

When Claremore City Attorney John Carle recommended that plans be tabled until further study could be done, my cup ran over. My temerity in venturing into the public arena had paid off; I didn't have to be a celebrity or a legislator to influence public events!

Eventually, the OCC approved the site as a "feasible location for an electricity generating station" but disavowed any jurisdiction over the kind of plant to be built.

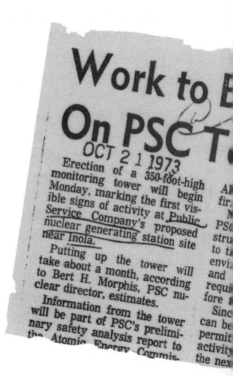

Work to E

On PSC T

OCT 2 1 1973

Erection of a 350-foot-high monitoring tower will begin Monday, marking the first visible signs of activity at Public Service Company's proposed nuclear generating station site near Inola.

Putting up the tower will take about a month, according to Bert H. Morphis, PSC nuclear director, estimates.

Information from the tower will be part of PSC's preliminary safety analysis report to the Atomic Energy Commis-

Our Paul Revere

It seemed that every time we turned on the television or opened up the newspaper something new was happening at Inola or about Inola. Before a construction permit could be issued, PSO would have to supply the AEC with a Preliminary Safety Analysis Report and an Environmental Impact Study, so they lost no time in beginning data collection. By mid-October, 1973, a 350-foot "weather monitoring tower" had been installed on the site. The tower held electronic devices that made a continuous record of such meteorological information as wind velocity and direction, air temperature at three different elevations, barometric pressure, rainfall, humidity and visibility.

The papers were filled with accounts and pictures of students from the University of Tulsa and OSU who had been hired to seine the river and count populations and species of fish and other marine life. Others tabulated mammal, lizard, snake, bird, and insect populations, while botany students were set to recording every species of plant life growing on the site, ranging from trees and grasses to mosses and lichens.

The installation of the monitoring tower was ominous. Once that happened, events began to gain momentum that would be very difficult to overcome. Once PSO had

Calif., firm. The pplies the tower. anticipates that cation for a con-mit is at least two s away. Both the l impact study ty analysis will erable time be- be completed. istruction work the site until a d, little other e place during months, Mor-

Environmental Aspects Eyed

Study Starts

Black Fox Station is the name Public Service Company has given its future nuclear power station near Inola. The company intends to build two 950,000-kilowatt nuclear generating units, one for initial operation in 1982 and the second in 1984.

Total cost of the plant, the first in Oklahoma, has been estimated at between $800 million and $1 billion.

Before this type of operation can go into effect, however, a lot of study and preparation must be carried.

While the definite site of the

Martin said. "We're exploring

Site Gets Tests

two years of weather data, a hearing on the construction permit would be held. Without the tower, there could be no data, no hearing, and no construction permit!

Where are our Paul Reveres? I asked myself. *If I were a man, I would go down to Inola and topple that tower!*

You may be asking yourself why a woman couldn't have done it just as well. Possibly she could, but for a woman of my generation, such a concept was—quite literally—unthinkable. The idea of my bringing down the tower never even occurred to me.

On George Washington's Birthday the next year, however, a Paul Revere did arise—but this time his name was Sam Lovejoy, and he lived in Montague, Massachusetts.

Northeast Utilities proposed to build two nuclear power reactors on the Montague sand flats bordering the beautiful Connecticut River in central Massachusetts. At the time, Sam Lovejoy was a young, politically and environmentally active farmer, who, with a group of friends, was engaged in the communal cultivation of a large farm in Montague. When the tower went up, Sam, like me, saw the handwriting on the wall. But unlike me, he did not pray for a rescuer. He took matters into his own hands.[1]

One morning, just before dawn, Sam walked the two miles from his home to the tower, climbed over the fence, and cut one of the huge cables that stabilized the tower. The tension on the cable was so great that he feared it would decapitate him in its thrashings, so he immediately dropped to the ground and quickly crawled away. The whir of the cable as it whipped back and forth seemed deafening, and he was sure people would arrive in moments to investigate.

By the time the cable had finally come to a stop, however, there was only silence. Not only were there no observers shouting at him to desist, there was no groan of a falling tower! The remaining cables still kept it upright. So Sam set about cutting a second cable. This time, Sam had to escape not only the lashing cable, but the sagging tower as well. The structure crumbled with a roaring crash, and this time Sam knew the whole countryside would be roused and come to arrest him. Still no dog barked, and no rooster crowed in the rising light of dawn.

Sam hitched a ride to town so he could turn himself in to the police.

"I destroyed their tower," he declared. "I sabotaged that outrageous symbol of a future nuclear power plant." But he went on to defend his actions: What he had destroyed was inanimate. The proposed nuclear power reactors, however, could destroy lives—including the life of his three-year-old daughter. Periodic radioactive releases from the facility into the atmosphere and the river could trigger leukemia or other lethal disorders caused by the free radicals that radiation produces in larger-than-natural quantities.

To be sure, the AEC hearings were supposed to protect the public from just such disasters. But Sam, like many other people, including me, felt that the hearings were a charade—a sham designed to hoodwink the public and lull them into a false sense of security. (Recent revelations have completely borne out our belief.) Preventing the hearings was, he felt, his only option, because it would prevent the construction of the facility.

Sam quoted the Massachusetts Bill of Rights: "The people alone have an incontestable, unalienable and indefeasible right to institute government and to reform, alter, or totally change the same when their protection, safety, prosperity and happiness require it."

"With the obvious danger of a nuclear power plant, with the biological finality of atomic radiation (and other equally ominous problems)," he told the court that arraigned him, "a clear duty was mine to secure for my community the welfare and safety that the government has not only refused to provide, but has conspired to destroy."

Again, quoting from the Massachusetts Bill of Rights, he declared that, "No man, nor corporation, nor association of men has any other title to obtain advantages, or particular and exclusive privileges, distinct from those of the community," and asked, "And yet, are we not now only beginning to grasp how grossly the great corporations view their profit?"

Using information he had gathered from public documents about a nuclear power plant in the nearby hill town of Rowe, Massachusetts—the oldest commercial plant in the country, and one whose safety record has often been cited by power companies to justify building others—Sam told the court that Rowe "had not been the impeccably safe place it had been so eagerly billed by the avaricious power companies; the plant had no emergency core cooling system at all until 1972! The ECCS," he explained, "is a rather simple water-cooling idea much like a car's radiator. It's supposed to control temperatures of several thousand degrees to prevent a meltdown."

(At the repeated urging of citizens' groups and federal and state legislators from the area, the NRC in 1992 denied Northeast Utilities' request to be allowed to continue to operate the Rowe plant despite evidence that radiation had fatally damaged the containment vessel. Upon investigation, it was discovered that the metal samples that were supposed to have provided an ongoing record of changes in the vessel had been discarded years earlier—or never put in place. Many people assume that Northeast Utilities believed, rightly at the time, that the AEC had no interest in safety, but rather in the appearance of safety, and discarded the samples as inconvenient—or possibly as damaging to their assertions of safety of the plant. The plant is currently being dismantled and its pieces were being shipped to nuclear dumps until space in the dumps ceased to be available.)

The community rallied to Sam's aid. Unlike the American Midwest, the northeastern part of the nation was dotted with nuclear power plants, and many people did not want any more built. Public hearings were convened and hundreds of people testified on Sam's behalf. Dr. John Gofman testified about the health effects of low-level radiation from nuclear power plants.

The courts refused to send Sam Lovejoy to prison, and to many people he became a kind of folk hero, a Robin Hood who went up against the powers that be—and won! I had personal reasons for being grateful to Sam. The Montague plant would have risen less than three miles from the home of my daughter, Patricia, and her husband, Edward Lemon; Sam had protected my Massachusetts grandchildren, Signe and Teddy, and the deed that I had only dreamed of he had accomplished! Eventually, Northeast Utilities' president declared that if the people of Western Massachusetts did not want the Montague reactors, NU would not build them.

PSO President R. O. Newman, too, once went on record to say that if the people of Oklahoma did not want nuclear power, PSO would not build the reactors. Yet as his company eliminated outdated and redundant jobs, the employees in those positions were trained to go into the community and speak at every opportunity, indoctrinating the public to ensure their support for a nuclear power facility at Inola.

During the fifties, one PSO home economist had come to my home economics classroom at Claremore High School to demonstrate the use and care of electrical appliances. Now she and her colleagues visited classrooms to promote the belief that nuclear power was the only way to "maintain the lifestyle we are enjoying now." Students and members of local organizations were informed that "nuclear power would provide more jobs as well as safe, clean, inexpensive energy."

Unlike the people of Montague, many in our area wanted nuclear power. At least they had been convinced they did! Perhaps the citizens of Claremore and Inola would have rallied around me if it had occurred to me to cut the Inola tower cables myself. After all, Westerners have a long tradition of individualism; many of our ancestors came west to escape the restrictions of the older society in the East.

However, Westerners also have a long tradition of respect for others' property, and PSO had convinced many of us that they owned the land and water and air of the site at Inola. Perhaps I would simply have languished in jail while PSO's plans went inexorably to their conclusion. Rightly or wrongly, I took the longer, more arduous and costly path of legal intervention. I set out to find a way of dispelling PSO's propaganda. I was only one person. They were many.

"And you work for PSO?"

Light Water Reactors Are too Dangerous
Says Britain's Leading Scientific Advisor

ECO section of Not Man Apart

ads the page one headline on *Weekly* ... *Report* (Washington, DC) June 10. ... the *Financial Times*, London, for June 7. ... is the letter that led to the *w*- ... *Report* story, a letter ...

ham [Vice Chancellor of Bath University], an Mr. Tombs [Chairman of the South of Scotland Electricity Board].

Water reactors can be of the pressure tube type, as in reactors such as CANDU and the

Atomic Industrial Forum Equates California Quest For Safe Nuclear Energy With a Moratorium

"In California, the initiative which would in effect create a moratorium on nuclear power has yet to qualify for the November ballot. The campaign to gather signatures to qualify the proposal does not seem to have attracted much attention, in contrast to previous attempts to create a moratorium in that state. The most visible action by nuclear critics has come from the Sierra Club, which asked its members to collect signatures for the petition."
                    --AIF Press INFO May, 1974

In the fall of 1973, I bought a film projector and a film entitled *What's Wrong with America's Reactors?* The film is a series of interviews with experts who explain that England had declined to buy America's nuclear reactors simply because they were unsafe. "How," it is asked, "if they weren't safe enough for the British, could they be safe enough for us?"

In one of the many prophetic utterances on the film, Professor Henry W. Kendall of the Union of Concerned Scientists talks about times when emergency shutdown mechanisms failed to operate properly and times when control of the reactor water level had been lost. "If by chance these two things should happen together (which is not impossible at all)," he says, "it could very well develop into an uncontrollable accident and a meltdown with catastrophic results."

Dr. Gofman talks about the major consequences of radiation, explaining that a meltdown could easily release as much as 350 units of radiation within a matter of hours, causing acute radiation sickness and a miserable death in a matter of weeks for half of the humans exposed. Trying to evacuate the population more quickly would be unproductive, he explains, because the most intense radioactivity would come early on. How, he demands, could we "get, say, several million people, out of London, inside of 12 hours...or New York."

Dr. Walter Jordon, then a member of the AEC's Reactor Safety Board, maintains that "Scientists and the public should be prepared to face the possibility of a nuclear incident just as we expect major earthquakes that will exact a large toll in property and lives," and that it makes sense to ask, "Have we succeeded in reducing the risk to a tolerable level?" (He explains that, to him and the AEC, a "tolerable" risk is one in which there is something less than one chance in ten thousand that a reactor will have a serious accident in a given year.) His question becomes nonsensical, however, when he admits that there is no way but experience to prove that any given level of risk is accurate.

Accepting one chance in ten thousand as acceptable risk would mean that if nuclear reactors proliferated as the AEC proposed, there would be, on the average, "a serious accident once every three years—losing a city for example." Dr. Gofman emphasizes that the AEC was not prepared to say that the risk was not even higher. In what I consider a most telling argument, he talks of the AEC's contention that "We just must get used to living with ... major nuclear disasters from nuclear power plants. Dr. Jordon," he notes, "says that's how we'll find out the risk of a major accident, and it's worth the risk. The real question is, do the inhabitants of London or New York or Philadelphia—who will be the experimental subjects in determining this risk—think it's worth finding out?"

For the next four years—until the NRC hearings began—I traveled throughout Oklahoma and into Arkansas, Missouri and Kansas, speaking and showing this film several times a week at schools and colleges and to civic and church groups—anywhere I could find a receptive audience.

Although most of the time I traveled alone, Robert or Mary would occasionally come along to help carry the projector and set it up, and when my journeys took me to Central and Western Oklahoma, Ilene Younghein often came along. Ilene is a delightful conversationalist, but, even more important, her technical precision and humor complemented my more down-to-earth, schoolmarmish talks. I felt the two of us, together, reached far more people than either of us could have alone.

After evening meetings, I would often stop in Oklahoma City to drop her off and end up staying the night and driving back the next morning. After a long evening spent talking to dozens of people—some friendly and willing to hear us, but others threatening and hostile—her home seemed like an oasis of peace and warmth. The blue willow plates and handmade rugs in Ilene's house made it as colorful and unassuming as its owner. With Ilene and Gaylord and their daughter, Gayle, I always felt at home.

Robert and Mary carried on with the operation of the nursing home. I was there some of every day, delegating duties and attending to patients, but to give myself more freedom of action, I hired two part-time registered nurses and sometimes called on our daughter Florence to help supervise the nursing care on her days off. I had two big jobs on my hands, and I wanted so much to do them both well, but without the willing cooperation of my family either would have been impossible.

PSO continued to watch my every move—just as I did theirs. Every time our meetings were announced to the public, PSO sent two people. We quickly came to recognize the PSO woman, but one of our principal amusements came to be guessing what the PSO man would look like. PSO apparently thought to deceive us by sending a different man as her escort each time!

In December of 1973, a man named Larry Bogart arrived in Tulsa wearing a shabby suit and cardboard insoles that were visible through the bottoms of his shoes. Ten years earlier, he had been a successful, impeccably dressed executive at Allied Chemical. That was before this Exeter and Harvard graduate came to realize the threat that nuclear power poses. Believing that it was his duty as a citizen as well as a human being, he had resigned from his job and begun a life devoted to awakening others. He founded the Citizens' Energy Council and for nine years had been publishing a monthly bulletin on the problems of nuclear power. He had spent his entire life savings trying to stop nuclear power, but when I asked him to come to Tulsa, he requested only that we buy his plane ticket.

He was a mesmerizing speaker—eloquent, well-informed, persuasive, and entertaining. The group that came to hear him sat rapt. As usual, the woman from PSO sat in the back of the room with that evening's escort and departed the moment the meeting adjourned.

Larry appeared on several Tulsa television stations and spoke on the radio before Robert and I drove with him to Ilene Younghein's house in Oklahoma City in preparation for the meetings and interviews she had scheduled for him. That trip tested our resolve, however, when it began to snow heavily, and all Robert's driving expertise was required to keep us alive in the treacherous road conditions! Hard as that trip was, it was even harder for us to part with this most remarkable, inspiring man. But the next morning we confided him to Ilene's care and set off for Claremore.

In January, of 1974, I saw an announcement in the *Tulsa World* inviting people opposed to Black Fox to a meeting. Someone else was trying to do something about the nuclear facility!

The "someone else" turned out to be three bright young men not long out of high school. Their names were Eddie Bryant, Kevin Chambers and Jay Nelson. Introducing myself, I asked if they would mind my showing the film after their presentation. To my surprise, they instead insisted that I conduct the meeting. And thus began a wonderful collaboration that was to be one of my major sources of moral—and practical—support in the years of our struggle with PSO.

At this meeting, too, the woman from PSO was accompanied by yet another man. Buoyed up by these newborn friendships, I felt full of mischief: "Let's introduce ourselves," I suggested as we began. "I'm Carrie Dickerson from Claremore. My husband and I own a nursing home, where I am in charge of nursing and patient care, but right now, I am an anti-nuclear activist." Then I asked someone opposite the PSO people to go next.

When their turn came, they gave their names but didn't tell where they worked. So I prompted them, "And you work for PSO?"

Sheepishly, they answered, "Yes."

I welcomed them, telling them they needed to see the film because nuclear injury is no respecter of persons. They could fall victim just as easily as anyone else. This time, they stayed to exchange pleasantries after the meeting was adjourned, but they knew they had been found out, and they never again showed their faces at one of our meetings.

14 A   FRIDAY, OCT. 3, 1975

studying 'safe sources'

Grandmother is nu

By KATIE RUFFIN
Of The Tribune Women's Staff

RRIE DICKERSON APPEARS TO just a helpless, mild-mannered andmother.
But there is muscle, she vows, and that soft voice and frequent ile — for she is organizing Tulsa a residents to oppose construc- n of a nuclear power plant near la and "in numbers there is wer."

Her group, Citizens' Action Group for Safe Energy Sources, has set u a booth in the Exposition building the 1975 Tulsa State Fair now through Sunday. Members are avail able from ... to 10 p.m... an swer ques tion and power.
" 'I' v thorough so dange

Nuclear
future e

By JIM EAST
O'Collegian Staff Writer

An opponent of nuclear faciliti foresee

A Hill of Beans

The spring of 1974 brought me a student from the University of Oklahoma who was writing a paper on nuclear power. He had used PSO people as a source of information but knew that his professor would expect him to present all sides of the issue, so he asked, "Is there any opposition to Black Fox?"

"They said, 'Only that little old woman in Claremore, but she doesn't amount to a hill of beans,' " he told me.

It struck my funny bone, so I laughed as I retorted, "But my hill of beans will grow like Jack's beanstalk."

That spring, too, there were two new nuclear projects under consideration for Oklahoma. The first was, ostensibly, a proposal for a coal gasification plant in the southeastern coal fields. Here coal was to be subjected to extreme heat to break it down into methane. It was only when one read in

THE TULSA TRIBUNE, TULSA, OKLAHOMA

...us of energy gro...

...op it," Carrie stated emphatically. "or the lives of future generations, have no other choice."

"But the compani... storage issue down an... wish," she said "IT...

...ER CRUSADE

...pponent sees ...rgy needs hurt

...itizens Action Group is ...ently circulating two petitions ...klahoma... Featured during the meetin...

the proposal, presented as an authorization bill to the Oklahoma State Legislature, that the heat required would be piped from a nearby nuclear power plant, that it became clear that it was really a backdoor proposal to build a nuclear plant! The second was a proposal by the Muskogee Chamber of Commerce to build a nuclear "park" on the lands of a former army base at Camp Gruber.

The Coal Gasification Bill needed clearance from the Oklahoma legislature. It had already passed the House, and Governor David Hall had been stumping the state talking it up in preparation for an impending vote in the state senate. I knew that I had very little time to find some way of bringing Oklahoma's senators' attention to the needs and will of the vast majority of Oklahoma's citizens. I went to Lewis Bean at Verdigris for advice. A former Oklahoma legislator himself, he told me how to get my message across—enlist great numbers of people in a concerted protest.

I wrote a four-page, single-spaced letter on legal-sized paper, devoting a page to each project. On the fourth page I listed the names, addresses and phone numbers of Governor Hall and the state senators and ended with a plea for everyone who was concerned about the well-being of future generations to write letters and make telephone calls. With this letter I hoped to enlist a great many people and turn their unified efforts into a weapon against the three projects. The major problem was, where would I get the names and addresses of the people who would become that weapon?

Let's Live and *Prevention* magazines had been publishing articles warning readers about the dangers of nuclear power, and I wondered if one of them might sell me its Oklahoma subscription list. Speed was imperative, and the people on that list would already understand what I was talking about. They would be faster to mobilize. So I decided to call Robert Rodale.

Robert Rodale and his father J.I.[1] had, unbeknownst to them, been honorary members of the Dickerson family for decades. My father-in-law, Eugene, had been one of the earliest subscribers to *Organic Gardening and Farming*, the first Rodale magazine, and "Rodale says ..." was often the authority of last resort on matters of ecologically sound farming practices. However, reaching Robert Rodale was a lot harder than calling up a brother. In fact, I think it would probably have been easier to get hold of the President of the United States. Finally, I reached him one Friday morning at breakfast. He told me there were over 18,000 Oklahoma subscribers to *Prevention* and that the list cost about $500. "But before I sell you the list," he told me, "I want to see your letter."

As Robert Rodale and I talked, my husband Robert sat across the desk from me in the nursing home office shaking his head and repeating, "I'll not spend another cent. Not another cent!" I pretended not to hear.

My Robert had been patient—extremely patient—understanding and generous. He had paid nearly $700 for the film projector and another $300 for the film. My car had an unslakeable thirst for gasoline, and my telephone bills were never less than $500 a month. On top of that was the cost of hiring additional nurses to cover my absences from the nursing home. I could understand his attitude: Enough was enough!

As I took the letter to the post office and went about my week-end work at the nursing home, I prayed silently, "God, Mr. Rodale has more money than I'll ever see. Please let him afford to give me the list!" On Monday morning, my prayers were answered: Robert Rodale called me to say that he was so impressed with my letter that he was sending the list immediately—and there would be no charge!

The Oklahoma Chapter of the Sierra Club gave me its membership list, and Friends of the Earth sent a list of its Oklahoma members. In all, I had almost 20,000 names. Those three lists made possible everything that followed.

I telephoned everyone in the area who had expressed concern about Black Fox and asked for help cutting and pasting addresses on the letters. College students, housewives, and school children manned the assembly line we set up in our house. Without a hint of a grumble, my dear husband wrote checks totaling $5000 to pay for printing and folding 20,000 copies of the letter and for the postage required to mail them. The message was sent on its way.

A tidal wave of letters and telephone calls deluged the governor and state senators. The coal gasification bill with its accompanying nuclear power plant was quickly defeated. One down and two to go!

Camp Gruber

In 1974, Americans were worrying about energy sources for the near future. These worries certainly had some real basis, but there is also reason to believe that, in part, they were manufactured in an on-going effort among utilities to expand their already indecent profits. As part of the search for short-range energy solutions, the AEC asked for and received Congressional authorization to identify appropriate sites for possible "nuclear parks" where related commercial nuclear facilities could be conveniently grouped. The Camp Gruber Military Reservation was under consideration for a series of demonstration projects, the first of which was to be a fuel enrichment plant to be followed by a fuel fabrication facility, several nuclear reactors, waste disposal facilities, and fuel reprocessing plants. In anticipation of this pork barrel the Muskogee Chamber of Commerce was ecstatic.

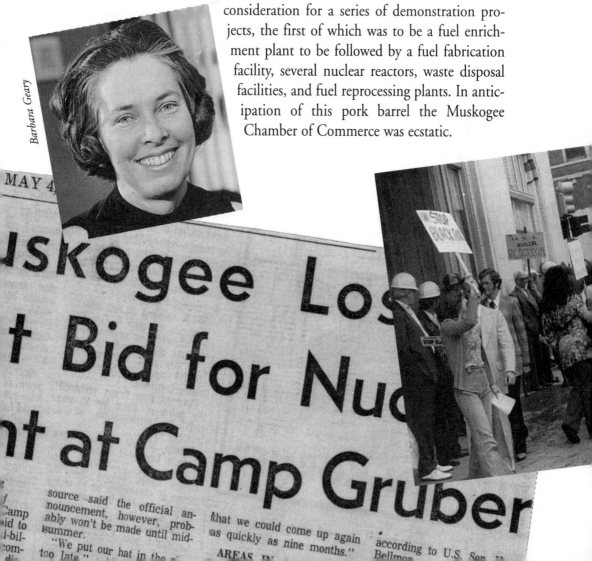

Barbara Geary

MAY 4

uskogee Los
t Bid for Nu
t at Camp Gruber

source said the official announcement, however, probably won't be made until midsummer.
"We put our hat in the...
too late."

...hat we could come up again as quickly as nine months."

AREAS IN

according to U.S. Sen...
Bellmon

In 1942, Camp Gruber was the site of a second Trail of Tears.[1] During the first one, the tribe was devastated by President Andrew Jackson's order in 1838, for the removal of the Cherokee Nation from its ancestral lands in Georgia, Tennessee, and the Carolinas to eastern Oklahoma. One hundred and three years later, in the single-mindedness of wartime, the forty-five Cherokee families living in the midst of thirty-two thousand acres of restricted Cherokee property adjoining Camp Gruber were given forty-five days to pack their belongings and abandon their homes.

In striking contrast to the humane treatment received by the prisoners of war (some were German soldiers from the Afrika Korps) who were housed on that land, the dispossessed Cherokees were paid grossly inadequate compensation for their houses and farms and received no relocation assistance of any sort. Even worse, it turned out that they had been forced to sacrifice everything for a camp that was deserted only three years later! A sensible person would have expected that the Cherokee families would then have had their land restored to them. A sensible person would have been wrong. By the time the AEC started eyeing it, Camp Gruber had been deserted for nearly thirty years.

Ilene Younghein's article, "Nuclear Park Questioned on Safety Grounds," was printed as a full-page story in Frosty Troy's *Oklahoma Observer*.[2] "Remember, it is better to stay up all night than to go to bed with a dragon," she warned.

That spring saw a number of meetings and symposia in Muskogee, Tahlequah, and other Oklahoma cities and towns—some sponsored by CASE and others by Northeastern State University and the Muskogee Chamber of Commerce. Their purpose was to give people an opportunity to consider and debate the implications—economic, environ-

Anti-nuclear protestors at PSO. Photo courtesy of *Tulsa Free Press*

Camp Gruber Energy Park Hearing Set

World's Own Service

MUSKOGEE — A public hearing on the proposed Camp Gruber Energy Park will be held here at 7:30 p.m. Tuesday in the Assembly Center here.

The hearing, sponsored by the Chamber of Commerce, is to determine the concerns of the public so that an upcoming study can concentrate on those areas.

Battelle Northwest Laboratories, Washington D.C., has been contracted by the Federal Energy Administration to study on the pro-mental, and social—of a nuclear park at Camp Gruber. Among the speakers were five professors from NSU at Tahlequah and U.S. Wildlife Agency fisheries biologist Jim Smith of Muskogee. Nuclear engineering consultant Richard Sandlin came from Washington, D.C., and University of Minnesota energy expert Dean Abrahamson and Oak Ridge (Tennessee) Laboratories Senior Site Representative David Cope spoke as well.

Defenders of nuclear facilities reduced the problems of nuclear parks to four points—producing adequate amounts of electricity and transmitting it to far-off cities, thermal pollution, freak accidents, and public acceptance—implying that they were all of the same nature as with any other energy source.

Opponents cited three major points—the cumulative and acute damage to life caused by radiation, the danger inherent in technology based on untested theories and relying on non-existent mechanical and human perfection, and the police-state mentality engendered by both greed and safety considerations. Eastern Oklahomans were also reminded that the thermal pollution of the rivers would destroy sport fishing—one of the major draws in the local tourism industry.

One speaker expressed his expert opinion that radiation from nuclear power plants "causes a life-shortening process"[3] that produces leukemia, cancer and birth defects over varying lengths of time. He also noted that because the plutonium such plants produce as a by-product is the main ingredient in nuclear bombs, it has an enormous potential for diversion to terrorist groups and nations. The necessary security precautions to protect ourselves encourage a police-state mentality that will ultimately destroy democracy.

At another symposium, Richard Sandlin reminded us that an experimental facility normally has the greatest potential for failure because its purpose is to test unproven theories. If the theories—and in many cases they were really only not-very-well-educated guesses—were wrong, humans and their environment could be severely damaged for hundreds of miles around. This was even more frightening in light of his claim that there were accidents "happening every day across the country in nuclear installations, ... accidents that, for the most part, were being effectively hushed up in an atmosphere more reminiscent of a police state than a democracy"—a claim now borne out by recent revelations from NRC records.

George Orwell, in his 1949 novel, *1984*, warned about the danger of state control over the dissemination of information. He might have been writing about the official secrecy and extensive public relations campaigns mounted during the development of the nuclear power program and continuing, to greater or lesser degrees, until today.

Many of these speakers held press conferences to spread their messages to the widest possible audience. Members of our organization testified at hearings in Muskogee. The Oklahoma Wildlife Federation urged a moratorium on promotion of the park. State organizations became polarized for and against.

On April 4, 1974, Robert and I were back in Stillwater showing the film and telling an audience of about 200 Sierra Club members and OSU students about Black Fox and Camp Gruber, when a pro-nuclear member of the audience took angry and vociferous exception to something I said. Before I could draw breath to voice a rebuttal, my usually taciturn husband faced the man and calmly explained why I was right.

To my surprise, a young stranger then rose and spoke vehemently in defense of my views. And thus I met Barbara Geary. Barbara, then a piano instructor at OSU, became my Stillwater liaison. Her classroom door was a bulletin board that educated her students about nuclear energy every time they came for a piano lesson. Barbara didn't just talk about the environment; she acted: To avoid adding pollutants to the atmosphere, she refused to own a car and, instead, rode her bicycle to work and the bus everywhere else!

When I mentioned the Nuclear Park Symposium to be held in Tahlequah on April 10, Barbara urged me to call her mother, Helen, who worked for the Corps of Engineers in Tulsa. Six days later, I picked Helen up at the rear entrance of the Corps building and drove with her to Tahlequah. In the years that followed, we held many meetings at Helen Geary's Tulsa home. After her retirement from the Corps of Engineers, I would think of Helen when I needed a place to rest between commitments. She tirelessly clipped articles for me—many of which I used to check out my memory of events for this book—and Black Fox's cancellation years later did not stop her. In the years after Robert died, when there was never enough money to subscribe to the *Tulsa World,* Helen Geary faithfully continued to keep me abreast of environmental and energy news.

In the wake of the non-violent protests triggered by Rosa Parks and used so successfully by Martin Luther King, demonstrations became probably the most popular avenue for arousing public outcry in this country. To be sure, Massachusetts Governor Peabody's elderly mother and Dr. Benjamin Spock had participated in such protests, but I thought of them as something for young people—kids who had not yet discovered a need to behave with dignity—so when I received a letter suggesting that we hold a protest rally in front of the downtown PSO office building to attract publicity, I was equally horrified and intrigued. Despite my qualms, it seemed like a good idea, so we made posters and obtained a city permit to hold the demonstration.

On the morning of the rally, I felt so ill that I went to my doctor. "What's bothering you?" he asked. "If your blood pressure stays this high, you're heading for a stroke."

Astonished because my blood pressure had theretofore always been subnormal, I confessed, "I can't bear it: Hundreds of my former students are going to see me on television, demonstrating like a flower child. They will think I have gone off my rocker!"

"Go ahead with the rally," was his kind advice. "You sit or stand quietly and talk with the people from the media. Let the kids do the parading."

My fears were in vain. While we were protesting, a former Inola student—totally unshocked by my radical behavior—came by, took some of our information, and visited a while. In fact, the whole day turned out to be great fun. Sudye Dalton and Cathy Coulson-Currin from the Sierra Club came with their children, who loved marching up and down the street with the posters. Helen Geary, Mildred Brown, and Betty Knight Broach were also prominent among those who paraded with the posters. We all handed out information sheets to passersby and stopped to talk with them.

To be sure, there were several tense moments when hecklers interfered with our activities. Opposite our position and down the street a ways, Tulsa police watched from two police cars. At the time, I wondered if they thought we might become violent and they'd have to arrest us, but now I suspect they were there to prevent heckling from turning into a riot.

As the day wore on, others joined in our protest. At one point a bus stopped in front of us, and a dozen or more young Native American men stepped off, picked up posters, and joined us. The one who attracted the most attention had his hair cut in a Mohawk. We learned that these young men had come to Tulsa from all over the U.S. to be participants in a national Native American conference.

The rally was a real success, attracting attention from all over the state, and reports were picked up even by the national media. While we may not have convinced everyone who heard about us, we made them think—and we gained many new supporters.

While we were intent on marshaling our forces, Senator Bellmon was doing his best to encourage companies with experience or interest in nuclear power to consider Camp Gruber. They didn't need much encouragement. Bechtel, Union Carbide, and Westinghouse were pondering a joint venture with private funds from both American and Japanese interests while the AEC dithered about whether the project should be a private or a public venture.

At a Muskogee meeting to discuss a site feasibility study, AEC officials warned that the project would cost many millions of dollars, that waste heat would raise the temperature of the Arkansas River by as much as 18 degrees, that fogging would be a problem, that radioactive tritium could get into the water supply, and that water use would be enormous. Then they told the 250 people there, "The time to start is now!"

Ilene and I urged that the money be spent, instead, on renewable energy facilities using solar and wind power to give us safe energy that would not saddle us with radioactive contaminants for millennia. We didn't quite know whether to laugh or cry when Congressman Risenhoover (a Democrat from Tahlequah) warned against "listening to these long-haired, wild-eyed radicals talking about what might happen if we have an energy center in Oklahoma," and ranted on that, "We can't afford to pay the Arabs

$12 a barrel for oil. We'll go bankrupt and we'll all be out of work."
However much Ilene and I stared at each other, our hair didn't get long,
and our eyes looked quite tame!

(June EchoHawk's were not: When we heard the same epithets at a
rally in Claremore, she stood up and said in no uncertain terms that being
long-haired and horrified about the prospect of Black Fox made her not a
radical, but a sensible person who had a right to wear her hair any way she
chose!)

Thomas Peace, then Director of the Oklahoma Department of
Pollution Control, argued that if the project were carried forth, "the State
of Oklahoma [might] have to sacrifice many existing and future water
needs, because almost all of the available water was already in use for
municipal drinking water, irrigation, industrial use, and hydroelectric
power generation. He also noted that parts of Camp Gruber were now
owned by a number of different state and federal groups, including the
Oklahoma Department of Wildlife Conservation and the U.S. Army
Corps of Engineers, and that the cost analysis that had been done did not
consider the cost to Oklahoma of losing recreational areas.

Despite its many drawbacks, U.S. Senator Henry Bellmon continued
tirelessly to promote the nuclear park. The Muskogee Chamber of
Commerce raised the money to charter a plane and take the dissenters and
those "on the fence" to Oak Ridge to let them see for themselves what a
boon the plant was to the area and show them that people were not dropping dead on the street because of the gaseous diffusion plant there.

On the appointed day, those who had been invited (among whom I
was not included) assembled at the Oklahoma City airport at 5 A.M. and
boarded the plane to await takeoff. Presently they were informed that there
was engine trouble, and everyone went home. As one of the most vocal
opponents of the park, I had felt snubbed to be left out, but the truth was
that I was relieved. I don't like to fly, and Oak Ridge was the last place in
the world I wanted to visit!

Perhaps they thought I possessed magical powers that had prevented
the scheduled take-off! More likely one of the original guests couldn't make
the rescheduled flight. Whatever the reason, I was invited to join the
reassembled party. This time the plane was boarded in Tulsa, and Senator

Bellmon sat across the aisle from me. We debated all the way to Oak Ridge and all the way back. As we toured the facility, he kept asking, "Don't you see, Carrie, that you're wrong? Don't you see that nuclear facilities are safe?"

Each time, I told him I'd seen nothing to prove me wrong.

Each time, he complained, "You're the stubbornest person I have ever met."

And each time I retorted, "I could say the same about you, Senator Bellmon."

At the research reactor, built from graphite blocks during the Manhattan Project, we stood by the railing and looked down into the beautiful turquoise glow from the uranium deep in its radiation-absorbing pool of water.

We climbed on a bus and went to see the mountain that had had half its substance ripped away to supply the coal that produced the electricity that operated Oak Ridge. We discovered that Oak Ridge had used more electricity than all the industries and homes of Nashville, Knoxville, and Chattanooga, combined. We learned that for every hundred watts of electricity the nuclear power plants in operation at the time produced, twenty-five watts went into the uranium enrichment operation.

(Oak Ridge sprawls over parts of two counties, Anderson and Roane. What we were *not* told during our tour was that for non-white women, lung cancer and leukemia mortality rates in those counties were four times what they should have been from 1950 to 1969 and that for twenty-two kinds of cancer, mortality in those counties—adjusted for age, race, and sex factors—was markedly higher. One can only conclude from that study[4] that Oak Ridge had been killing the surrounding population slowly, beginning with the non-white females. If there had been an individual perpetrator of those atrocities, that person would have been tried, found guilty and executed. And yet nuclear power was being hailed as the salvation of mankind!)

When the plane landed in Tulsa, Senator Bellmon stood up and faced me. Once more he repeated, "Carrie, you're the stubbornest person I have ever met."

"Senator Bellmon," I held his eye. "I've given you a whole day of my life, and you have not shown me one thing that could change my mind.

You owe me a day. I'm inviting you and your wife to spend a day at our farm with my husband and me. We'll convince you that you're wrong."

By this time people were clamoring to leave the plane, and we were blocking the way. In chorus they shouted, "Give her a day, Henry! Give her a day!"

That was the last time I ever saw Senator Bellmon in person.

Camp Gruber was not the only site being considered for the experimental facility. Union Carbide and Westinghouse pulled out of the partnership; Bechtel finally found another partner, Goodyear, and decided to build a gaseous-diffusion plant near Dothan, Alabama. That was small consolation, because more fuel for nuclear power plants—no matter where it was produced—would only ensure the building of more nuclear power plants.

For a while Senator Bellmon and the Muskogee Chamber of Commerce continued to promote a second such plant at Camp Gruber, but eventually everyone concerned came to agree with Mr. Peace that it was not feasible, and the project was abandoned in 1975.

Two down and one to go! How jubilant we were! And how happily oblivious to the true dimensions of what was to come.

The Debate Continues: 1974-75

Nuclear energy was being increasingly touted by the AEC as a cheap, reliable source of power, but informed citizens were constantly on the watch for misinformation, disinformation, and omissions that allowed erroneous implications to stand. Commissioner William Doub, in a letter to the editor of the *Saturday Review*,[1] claimed that currently operating nuclear reactors had a capacity equivalent to 300 million barrels of oil per year. His implication, of course, was that all of that capacity—about 20,000 watts—was available for consumption.

In September, the magazine printed a rebuttal[2] by Washington, D.C., lawyer Anthony Z. Roisman, in which he pointed out that, in practice, currently operating plants were delivering only about 60% of their theoretical capacity; thus the actual capacity was only about 12,000 watts. This meant that construction cost per kilowatt hour, instead of $600 million, was actually $1 billion.

"Before we are irrevocably committed to the super-dangerous nuclear technology," he urged, the country should make the much lesser investment required to "carefully

Anthony Z. Roisman

A Strange Beast, the JCAE; It Calls All the Nuclear Sho

There's nothing quite like the Joint Committee on Atomic Energy (JCAE). The JCAE, made up of members from both the House and handles all bills dealing nuclear power. Because ment usually given the E's members of nuclear

Third, the JCAE is the only joint committee which can authorize the spending of appropriated funds. Fourth, the requirement th the AEC keep the committee and continually informed in the JCAE in the day-to-day w of the commission, tiein together. Instead of rec AEC, the JCAE has c the commission to co tion on nuclear da to Ralph Nader. hand

examine … alternative energy-supply systems, such as solar, wind, geothermal, sewage and solid waste … With a rational program of energy conservation," he contended, "we could abandon the nuclear reactor program and be independent and safe by 1980."

Recent revelations about members of Congress suggest that in many cases senators and congressmen on powerful committees have concealed or ignored their own serious conflicts of interest, including direct ownership of stocks or bonds that are affected by the legislation they oversee or such ownership by relatives and supporters or large campaign contributions (in money and in favors) by businesses affected by their committees. I do not know whether that was the case with members of the Congressional Joint Committee on Atomic Energy or whether they simply had to close their eyes to the dangers of nuclear power to salve their consciences. Whatever the reason, the JCAE had for years, in concert with the AEC, promoted nuclear power and ignored its dangers. By and large, in fact, the JCAE and the AEC spoke with one voice.

They did not have the excuse that they had not been told how monstrous their actions were: In early 1974, Ralph Nader told the JCAE that the AEC had been concealing a staff report that made it clear that the latter had for years been conspiring to deceive the public about reactor safety. That report indicated that the reactor safety statistics frequently used in nuclear power promotion were invalid and that safety problems were "besieging" nuclear plants. Moreover, need for new, more stringent plant safety requirements, had also been concealed on the grounds that revealing them would frighten the people living near the ten currently operating plants that failed to meet the new guidelines. The JCAE ignored Nader.

Instead, the committee began considering ways to further restrict the public's already limited right to participate in nuclear power decisions, thus shortening the time it took to plan and build a nuclear power plant. It seemed that on the national front the cards were stacked against opponents of nuclear power.

Yet other Congressmen gave us hope. Senator Mike Gravel (D, Alaska) introduced a nuclear power moratorium bill in Congress. The Washington, D.C., Task Force Against Nuclear Pollution sponsored a national petition drive in support of the moratorium, and CASE members gathered signatures at rallies, fairs, and anywhere else they could find people. Even when I went to the grocery store, I would take my petition with me and stop anyone who would listen as I went down the aisles; I nearly always left with a new set of signatures.

Signature gathering worked in two directions: It helped us get our voices to Congress, and it helped us get our message out to more and more people. Sierra Club members helped collect signatures at the Tulsa State Fair each fall, and one of those fairs was the scene of a very illuminating experience. A woman who had already signed our petition later returned with her brother. When she asked him to sign, he became very angry and let fly a volley of obscenities. When he finally stopped for breath, I seized my opportunity. "How much do you own in uranium royalties?" I asked gently.

"How did you know?" he demanded in belligerent surprise.

"I'm not a mind reader," I assured him, "but you were so defensive I knew you must be protecting something, and I guessed it was the money you were making on uranium royalties."

By 1974, the wartime and McCarthy-era willingness of the public to give *carte blanche* to the Atomic Energy Commission was a thing of the past. Citizens again felt confident in demanding that government be answerable to the people—rather than the other way around—just as the Constitution provides. In this atmosphere, the worst excesses of the AEC came to be recognized as natural given the fact that the agency had responsibility not only to regulate the nuclear power industry but also to promote it.

Thus when in 1975 the House of Representatives, by a vote of 372 to 1, passed a bill to institute a $20 billion, 10-year search for energy self-sufficiency, its provisions reflected this renewed understanding of government in a democracy. Passed by the Senate and signed into law by President Ford, it abolished the AEC and divided its responsibilities between two new entities. The new law directed one of its successor agencies, the Energy Research and Development Administration, to explore all available avenues—including renewable sources—rather than nuclear energy alone. The other, the Nuclear Regulatory Commission, was directed to limit itself to regulation.

A civilian, nominated by the President and confirmed by the Senate, was to head the ERDA as it converted the AEC's $10 billion complex of national laboratories to more universal research in energy technologies. The five members of the new NRC were also to be subject to presidential nomination and senate confirmation.

By the spring of 1974, leaders of the National Farmers' Union and the Kansas Farmers' Union had come out in opposition to nuclear power. Robert and I hoped that Oklahoma leaders and members of the organization that had provided the backdrop for our first acquaintance and in which we had worked so hard over the years would support our cause. Then in Oklahoma City we learned a lesson that anyone involved in change must take to heart: *The ground must be prepared in advance if change is to take place successfully.*

We went to Oklahoma City to the annual Oklahoma Farmers' Union Convention, where we renewed friendships and acquaintances and talked with people we had never before met, handing out leaflets and explaining the dangers of nuclear power to most of those at the session. Then Robert introduced a resolution opposing the use of nuclear power in Oklahoma.

Before it came to the floor, the Resolutions Committee had revised Robert's text completely. The Tennessee Valley Authority was then initiating the use of nuclear power by its public utilities, and by the time the committee got through with Robert's resolution, its wording supported the idea that Oklahoma should follow the TVA's example!

We felt betrayed, but we refused to turn tail and run. During the general session Robert attempted to amend the committee version to reflect his original intent and was scolded by the OFU President. "If my old friend Gene were here," this man blustered, "he would be shamed, shamed, I tell you, to see his son joining the radicals that are trying to destroy Oklahoma's future."

It didn't matter that Robert's father, a long-time member of the state and national boards of the Farmers' Union, in fact supported his son's stance. It was beside the point that "old friend Gene" knew the nuclear beast for what it was—an outlaw that could mean painful illness and death to his present and future descendants. No one would ever know it, because the session chairman ruled Robert out of order before he could tell them so. Robert and I felt so disappointed and humiliated that we left the meeting and went home.

In late October of 1977, the Oklahoma Legislative Interim Committee on Environmental Quality held a hearing on the Black Fox nuclear power plant. Here, government officials also had disquieting information to share with the committee.

Among them was Robert L. Craig, Director of the Oklahoma Health Department's Radiation Detection Division. He was even-handed in his approach, noting that while nuclear plants pose problems serious enough that "the health department [had] asked for a $102,000 [budget] increase to finance monitoring of the Black Fox plant," other forms of energy production also damage the environment. He did not feel the need to mention that the $102,000 outlay would increase with time. Most taxpayers were not aware of the extra state tax dollars Black Fox would cost, but Craig knew the committee members were not so naive.

One committee member, Representative Gary Payne (D, Atoka), was concerned about the "potentially grave" problems associated with nuclear power. He urged the legislature to "write laws covering locations, size of plants, security of plants, and waste disposal."

CASE paid the way for Albuquerque environmental physicist Dr. Charles Hyder to appear as an expert witness at the hearing. Dr. Hyder told the committee that there are many possible initiators of nuclear power plant accidents—including human error, equipment failure, earthquakes, severe storms, and "acts of hostile humans."

Lies by omission became part of the testimony when PSO's James Parmley and Joe Bevis admitted that "materials become highly radioactive after the fission process" but implied that that was no problem, because wastes from Black Fox would be stored not in Inola, but at a reprocessing plant. What they did not say was that at that time the only plant in existence in the U.S. for processing radioactive wastes[3] from civilian nuclear reactors had contaminated the environment so terribly that it had been closed permanently in 1972.

When these men were questioned about permanent disposal of the wastes, they gave legislators the standard industry response—that it was not yet planned. Understandably, they did not mention that it was not yet planned because no one had yet come up with a satisfactory solution to this problem!

At the close of the interim study, Oklahoma House Environmental Quality Committee Chairman, Thomas Bamberger (D, Oklahoma City) declared himself personally opposed to nuclear power plants and other facilities using radioactive materials. That spring, he introduced two resolutions in the legislature, one calling for a state referendum on a proposal to ban the "use, manufacture or storage of plutonium" in Oklahoma and the other to "authorize the Legislature to regulate the location, number and operation of nuclear processing facilities and nuclear power plants." He told reporters that he did not want to be responsible for "burdening untold generations ... so that we can have abundant power [today]."

Hopeful as this sounded, we knew that the NRC would not give up so easily. They might have had a new name and a new mandate, but old habits do not die so easily: They were still in the business of promoting nuclear energy—and they remain in that business, today, despite all their disclaimers. The only difference is that now they speak through the companies they supposedly regulate.

We were right not to hope too much: The NRC maintained that the states had no authority to regulate the use of nuclear materials, and both resolutions failed in legislative committee.

Death o
Worke

CRESCENT (UPI)—
form-minded Karen Sil
auto crash, Kerr-McG
ting down its giant nu
the Cimarron River.
Miss Silkwood's co
into a concrete culv
She was enroute to t
ficial and a New Y
about alleged saf
plant.
Miss Silkwood, 2

2 SECTION H

Karen Silkwood 'Fe|

OKLAHOMA CITY (AP) — Karen Silkwood felt she was going to die, but she was determined to prove her claims the Kerr-McGee Corp. falsified its nuclear fuel rod record.

McGee attorney during the deposition, Mrs. Jung gave a detailed description of the notebook and file folder Miss Silkwood

with The
mission,
to a law

Critical Mass '74 and Karen Silkwood

For years the announced goal of the government had been to make the United States dependent upon nuclear energy by the year 2000. By the second half of 1974, they had succeeded so well that forty-nine nuclear reactors were already in operation, fifty-eight were in the process of being built, and 122 others were on order or were in the proposal stage.

The tendency of most Americans, fostered by the Depression and New Deal and by World War II, had been to believe that even those agencies that were most secretive were run by patriots who believed in democracy and who would

SECTION F 13

err-McGee
fected Mar

plant a year ago, said the
reduction of employes from a
270 had little effect on his town
It hurt Oklahoma City, about
utes away, and nearby town
because most workers commu
said.
"We lost a few school child
as far as businesswise we do
like it's hurting," he said. "W
a small town and every persor
helps us. That's one reason
went to bat for them."

RALPH NADER

A WORLD, T

Death Near'

against the ad-   cafe in Crescent, Okla. near th
ony. According

Above Left: John W. Gofman, M.D., Ph.D.
Above Right: Ralph Nader at Critical Mass

thus be working for the good of all. By the seventies, however, there was much evidence that autocratic groups ultimately work only for their own aggrandizement. People were beginning once more to realize the truth of the Thomas Jefferson's dictum, "Eternal vigilance is the price of liberty," and to rouse themselves to scrutinize such groups. When they did, they didn't like what they saw. They started organizing themselves to take back their liberty. Ralph Nader was one of the most vocal—and certainly the most famous—of those who began to speak out.

By this time several national groups organized to combat the nuclear threat had become well known. Among them were the Union of Concerned Scientists, Friends of the Earth, National Task Force Against Nuclear Pollution, National Intervenors, National Resources Defense Council, and the Committee for Nuclear Responsibility. Hundreds of lesser-known groups across the country were also struggling with the nuclear giant, and among these was CASE.

In the middle of November, 1974, Nader sponsored a conference at which he hoped to assemble the fragmented grassroots campaigns, together with the national groups, into a nationwide crusade for a nuclear power moratorium. Its theme was "A nuclear catastrophe is too big a price to pay for our electric bill." Because it was premised on the notion that after a public opinion crusade reaches a certain level, it is unstoppable—just like a nuclear chain reaction—he called the conference Critical Mass. The gathering was designed to collect the separate activist groups in one place and provide a national focal point for discussion of the risks and consequences of nuclear fission, thus bringing about a critical public-opinion mass.

I flew to Washington on November 15, along with more than a thousand other members of grassroots groups. Ready to share what I had learned, but eager to learn new and more effective techniques to use in our efforts to stop Black Fox, I went off to the first session at the Statler Hilton. The conference had also attracted many scientists and members of Congress along with a few celebrities. I met dozens of people from across the nation who had already shared their experiences and insights with me by letter and telephone. I became better acquainted with them as we attended workshops and listened to speakers who were convinced that atomic fission power was one of the most dangerous threats to the survival of the human race.

At the opening session, a Congressional panel comprising Senators Mike Gravel (D, Alaska) and William Proxmire (D, Wisconsin) and several members of the House of Representatives listened to eight prominent citizens summarize our major concerns about nuclear power—and its chilling effect on discourse in a democracy. These people were Nobel Prize-winning physicist Hannes Alfven, former Pennsylvania Insurance Commissioner Dr. Herbert S. Denenberg, Union of Concerned Scientists engineer Carl J. Hocevar (a seven-year veteran of the AEC Safety Research Center), Dr. John Gofman, Another Mother for Peace Executive Director Dorothy Jones, University of Rhode Island Professor of Regulatory Economy Richard Hellman, North Anna (Virginia) Environmental Coalition Treasurer June Allen, and the Reverend Albert Fritsch, Director of the Center for Science in the Public Interest.

The burden of their testimony was that citizens cannot abdicate responsibility in favor of scientists and career civil servants. They reasoned that scientists and civil servants often have a vested interest—personal, professional, or financial—in the success of a particular initiative and also that as human beings they are not above acting for their short-term personal good instead of the long-term public good. This was the only reason, Carl Hocevar believed, that evaluations of safety risks relying on untested assumptions chosen to support a given outcome, as in the Rasmussen Report, were given any credence at all.

Dr. Denenberg explained that the Price-Anderson Act did not give any insurance protection to the consumer. Dr. Gofman spoke for most of the anti-nuclear hosts when he said, "The essence of the problem is exceedingly simple—if we generate nuclear power to meet any significant proportion of our energy use, we create astronomical quantities of radioactive fission products and plutonium[239] ... the problem becomes whether or not these substances can be perfectly isolated from the biological environment almost forever." From Richard Hellman, we heard an exposé of the true costs to the public of nuclear power, and June Allen told of engineers so careless of the public welfare that they concealed the existence of a geologic fault line running under four reactor sites.

I packed as many workshops and seminars into my weekend as I could manage. The first was on the mechanics of utility regulation. Like most people at the time, I was very naïve about this issue. In my youth, I had heard vaguely about the excesses of the railroads in the nineteenth century, but I thought the Sherman Anti-Trust Act had taken care of that. Where power utilities were concerned, we had relied on coal oil lamps for light and draft horses and gasoline engines for power until after World War II. Then our municipal utility, Claremore Light and Water, extended its power lines to the surrounding countryside.

Robert had bought a "do-it-yourself" book and learned how to install electricity in our houses. Ceiling lights were installed, switches and outlets put neatly in place—and the extra holes plastered over and painted. The day the electricity was turned on in the house was a day for celebration. No more lamps to be filled with smelly coal oil. No more lampblack to clean off the chimneys. No more wicks to be trimmed or mantles to be replaced. No more strained eyes from trying to sew or read with inadequate light! I don't know whether I would have cared then had I known just how few teeth the utility regulations contained.

Other workshops taught me just how adjustment clauses encourage the building of nuclear power plants, how rates are structured, how to get the state to become a co-intervenor, and how an intervenor might go about convincing the AEC hearing board that they are being asked to approve a plant whose services the public can't afford. Still others gave us an opportunity to listen to experts in fund raising, intervention, accidents, worker safety and radiation hazards, and the economics of nuclear power.

At the seminar on signature gathering, I met Franklin Gage, a young man from the National Task Force Against Nuclear Pollution. CASE had helped him collect signatures on a nuclear moratorium petition he had initiated. So far, he had collected 115,000 signatures toward his goal of one million, a number that he hoped would impress Congress enough that they would pass Mike Gravel's moratorium bill.

Dr. Ralph Lapp, an energy industry consultant, and William O. Daub, a former Atomic Energy Commissioner, presented pro-nuclear points of view and then responded to questions from the floor. I stood up. "Dr. Lapp," I told him, "just two evenings ago Karen Silkwood, an employee of a plutonium plant in Oklahoma, was killed in an automobile accident. Previously, on three occasions she had been physically contaminated with plutonium. Through unknown means her apartment had also been contaminated with plutonium. If anyone could take that plutonium off the plant grounds without detection," I asked, "doesn't that prove that no safeguards are adequate to protect the public against its dangers?"

Kerr-McGee and Karen Silkwood made international headlines when on the evening of November 13, 1974, Karen met her death. Her car crashed into a culvert as she was on the way to Oklahoma City. Before her death, Karen had been a worker in Kerr-McGee's Cimarron plutonium processing plant at Crescent, Oklahoma. The plant produced plutonium 239 fuel rods for the Fast Flux Research Reactor, the experimental liquid-metal fast breeder reactor in Hanford, Washington, that President Nixon had wanted.

On September 27, 1974, seven weeks before her death, Karen had gone to Washington, D.C., with two of her fellow workers, to meet with AEC officials. They accused Kerr-McGee officials of falsifying fuel-rod quality-control and safety records. Karen had evidence that, she believed, showed that a worker had used a felt tip marker to alter x-rays of faulty welds to make them appear sound. If she were right, Karen feared, the faults would lead to an accident that would release huge amounts of radioactivity.

Karen and her co-workers further charged that plant officials consistently and unnecessarily endangered the lives of workers. At the most basic level, they told AEC officials, Kerr-McGee left them ill-trained in proper procedures for handling plutonium. Furthermore, they believed the company avoided educating workers about the potential dangers inherent in their jobs. They believed this made it possible for the company to ignore excessive worker exposures to radioactivity and to fail to monitor worker exposure adequately or take appropriate hygienic precautions.

Five weeks later, a plant accident exposed Karen to large amounts of radiation and contaminated her body with plutonium. A week after that, plutonium-contaminated cheese and bologna were found in her refrigerator, and additional contamination was found in her bathroom. Everything in her apartment—including all her mementos and photographs of her children—was wrapped in heavy plastic and shipped to a special nuclear waste burial ground. (Karen was a divorcee, and—fortunately for them—her three young children lived with her parents while she worked.)

This time, Karen herself was sent to Los Alamos, New Mexico, where her lungs were washed out in an attempt to remove the plutonium, and her body was tested for other plutonium contamination. Medical workers reported that her body contained 2000% of the maximum allowed under AEC safety guidelines. Why did Kerr-McGee's security not prevent plutonium from leaving the plant? And how can we be sure that the next time, it would not have left the plant in the hands of terrorists, or their venal suppliers?

A week after her return from Los Alamos, Karen died en route to a meeting with a labor union official and a *New York Times* reporter. The purpose of the meeting had been for her to give them documentation of the unsafe practices on which she had earlier blown the whistle. The accident was not discovered until the next morning, and by then the documents were no longer in her car. Nothing else was missing. They have never been recovered. Someone didn't wish to be found out. It is unfortunate that Karen allowed others to know of her plans. Had she not, perhaps she would be alive today.

Sadly, Karen's death attracted far more attention than any of the warnings she had tried to convey about what the Cimarron plant was secretly doing to the public and to its own employees. The Oklahoma State Legislature began to realize that they had to do something to calm public fears, and on January 30, 1975, they opened a formal investigation.

My son James accompanied me to Oklahoma City, where we listened as Kerr-McGee plant officials underwent a lengthy interrogation. The investigation ended—as do all those whose real purpose is to soothe voters and give legislators a public grandstand—with the committee exonerating the company and its officials of any wrongdoing. The committee had, however, misread the temper of public opinion. In fact, their white-washing efforts only added fuel to the fire.

When the plutonium fuel rods did arrive in Richland, Washington, reexamination showed that about 400 of them were, indeed, faulty and could have caused a disaster if they had actually been used. Eventually, Karen's death led to the publication of a book[1] and a movie, *Silkwood*, starring Cher, Meryl Streep and Kurt Russell. Karen gave her life to prevent a nuclear holocaust. The evidence she uncovered was directly responsible for the closing of the Kerr-McGee Cimarron plant in 1975 and contributed to the later closing of the Fast Flux Research Reactor.

Evidently, the people at Critical Mass '74 had not yet learned of Karen's death when I spoke. I knew because Ilene had told me when I called her that morning. Many people came up to me and shared their concerns and their sadness that her life was lost because she had been trying to improve safety conditions and workmanship at the plant. Dr. Gofman was one of them.

When I told him of my plans to intervene in the NRC hearings on Black Fox, he advised me not to do so, telling me that I would be wasting my time, my energy and my money. He had testified at numerous hearings and found them to be nothing more than kangaroo courts. His testimony had fallen on deaf ears.

Another panel discussed the dangers of plutonium and talked about purported NRC plans to recycle plutonium as fuel for light water reactors. Here was a new worry: Would Black Fox use plutonium for fuel?

When anthropologist Margaret Mead stood up—old and frail and leaning on a tall, crooked shepherd's staff—it seemed as if a beloved monument had come to life. Through all the Black Fox skirmishes and battles, one of her *dicta* was a continuing source of inspiration to me: "Never doubt that a small group of thoughtful, committed citizens can change the world. Indeed, it is the only thing that ever has."

But it remained for chemist Dr. Chauncy Kepford to bring forth a wry chuckle when he opined that the reason there hadn't been much government support or industry interest in solar power was that "the utilities haven't been able to figure out a way to monopolize the sun!"

Others who spoke included Leo Goodman, formerly of the Auto Workers union and one of the first active veterans of the anti-nuclear movement, and physicist Henry Kendall and Executive Director Daniel Ford of the Union of Concerned Scientists. Two of the speakers, Professor Dean Abrahamson of the University of Minnesota and Charles Komanoff of the Council on Economic Priorities, later acted as expert witnesses for us. Actor Robert Redford came to listen and learn, along with biologist George Wald of Harvard, who later won a Nobel Prize.

It was a wonderful feeling to realize that I was a part of a growing mass movement that was coalescing from our country's anti-nuclear sentiment. That conference filled me with renewed hope, faith, enthusiasm, and empowerment and reaffirmed my resolve that we must stop not only Black Fox, but all nuclear power!

TURN ON WITH SOLAR ENERGY

36—STILLWATER (OKLA.) NEWS-PRESS—Wednesday, June 25,

Energy Solution Blowing In Win

TOPEKA, Kan. (AP) — The answer to this nation's energy problems may be literally blowing in the wind, a couple of ___ ___er Kansas assistant ___ ___lieve.

for Windustries.
Joiners have come from virtually all sections of this country and from Canada.
Burr and Ward both are involved in the work of The Villages, a Topeka project in___ ___ by the Menninger Foun___ ___ counseling

He said ___ hopes to win ___ veloping sol ___ the nuclear ___ ceived in ___
"The t ___ here," h ___ moon in ___ develop ___ nology

Citizens Action Group Becomes CASE

Critical Mass '74 opened my eyes more fully to some personal issues: Our most active members had become increasingly concerned about what would happen to our families if the organization became legally liable for a large amount of money. As an unincorporated group, we would be individually, rather than collectively responsible; our families could suffer. Others argued that we would be taken more seriously if we were incorporated. The more we talked about it, the more necessary incorporation seemed.

However, there were several things that needed to be taken care of first: We looked at our name and decided that we wanted one that had an easily remembered acronym. We finally settled on "Citizens' Action for Safe Energy," which said very clearly what we were about and which had the easily remembered acronym CASE. We needed articles of incorporation, bylaws, and a statement of purpose. And we needed a board of directors.

Several of us went to Oklahoma City for the Karen Silkwood hearing. It seemed a perfect opportunity to do the paperwork, so during a recess in the investigation on January 30, 1975, five of us—Ilene Younghein and Doris Gunn from Oklahoma City, J. Kevin

Chambers from Tulsa, and my son James and I from Claremore—acted as temporary board members until we could hold an election. We went to the office of the Secretary of State and signed articles of incorporation for Citizens' Action for Safe Energy as an educational and charitable organization.

The next thing we needed was a logo. Since we wanted to tell people that we had a positive agenda—the promotion of safe, renewable energy sources in place of suicidally dangerous ones—I drew a scene with the sun, a windmill, a waterwheel, and animals (representing solar, wind, and water power and methane) for our stationery. I painted the same scene on an apron and wore my renewable energy message wherever I went. Soon members began holding apron-painting parties as part of our fund-raising efforts.

Cathy Coulson-Currin painted one of her delightful watercolors for us, a red fox with the slogan, "The Red Fox, the Good Fox. Stop Black Fox!" written below. Later, she added to her portfolio a gray fox and then two fox kits. Reproductions of these paintings also helped us raise funds.

As time went on, we added to our arsenal of fund-raising items: T-shirts silk-screened with the red fox or a sun-and-windmill motif on frames built for us by Gary Laner; buttons with the slogan, "Solar energy turns me on!" surrounding a smiling sun; sun-and-windmill pendants; and various bumper stickers. Contributors received something they could enjoy and at the same time helped spread the message farther.

On March 31, 1975, a worker at Browns Ferry, a TVA reactor near Athens, Alabama, was looking for air leaks in the electric cable room. Reasoning that an air current would bend a flame, he carried with him a lighted candle. Unfortunately, he either did not know the cable coatings were flammable or did not understand the implications of that fact. When his candle flame ignited the coating, the fire spread quickly, burning the insulation from about 1600 electrical cables. They short-circuited, putting out of commission controls for valves, pumps and blowers as well as data-collecting instruments that were supposed to feed information to operators in the control room.

Without power, the emergency core cooling system for the Unit 1 reactor and portions of Unit 2 could not be operated. In the darkened, smoke-filled room, the desperate reactor operators had the presence of mind to shut down both reactors—by hand—in the nick of time and to bring in hand pumps to pump cold water into the reactor cores, thereby saving millions of lives. Many hours after the fire began, it was finally extinguished—leaving behind about $500 million in damage.

Despite what one would think was an instructive experience, TVA engineers installed new support structures for an enormous reactor pressure vessel sideways a year later! Demonstrating that they were better at jury-rigged band-aids for emergencies than at preventing them, the engineers installed shims to correct the problem. That would have been funny, were it not so frightening. And TVA and the NRC didn't want the public to be frightened: If we all truly realized how much depended on human accuracy and people who can think in an emergency, we might refuse to let the utilities have even more dangerous and expensive toys.

Little by little, the understanding that nuclear power was not a cozy, harmless housecat, but rather a man-eating tiger, began to seep into the consciousness of people around the country. When the Oklahoma House Environmental Quality Committee held its hearing on possible state control of nuclear plants, OU chemistry professor John Burr warned them that, by the very nature of fission, nuclear reactors can be expected to be subject to chronic breakdowns, large and small. He asked, do we have the right to "create radioactive poisoning which will have to be guarded in the environment for thousands of generations?"

Professor Connie Taylor, who taught biology at Southeastern State University in Durant and chaired the Oklahoma Sierra Club Conservation Committee, took the opportunity to explain that, "What may appear to be low levels of radiation concentration in soil, air or water is deceiving." She explained how, through the food chain, cesium[137] so dilute as to be almost negligible when it leaves a power station, becomes lethally concentrated in sunfish and carp and how strontium[90], chemically similar to calcium but with a radioactive half-life of thirty years, becomes concentrated in our bones—including the marrow where blood cells are produced—causing leukemia.

Her final point could not be other than sobering to everyone in the room: Leukemia (blood-cell cancer), which hardly existed in the first half of the century, has begun to reach "epidemic proportions, particularly in children." The timing of this epidemic is very suggestive: It began in the sixties, about the time the results of fallout from atom-bomb testing would have had time to begin appearing.

On November 12 and 13, 1975, Dr. Taylor expanded on this theme when she told the Claremore and Catoosa "Issues and Values 1975" seminars that, "Any amount of radiation, no matter how small, is bound to produce some mutations, that mutations are irreversible and almost always harmful." (These widely publicized and well attended seminars were part of a series held all over Oklahoma; my daughter Mary and I provided refreshments for the approximately seventy participants at each event.)

Certainly the committee members found the testimony they heard convincing. They produced two resolutions, one banning nuclear power plants in the State of Oklahoma, the other allowing Oklahoma voters to decide whether the legislature should have the power to regulate nuclear power plants in the state.

Even though state representatives as a whole voted 77-14 against the two resolutions, we felt that the attempt had been worthwhile. It provided both a public forum and a great deal of publicity. Because the legislature is thought of in Oklahoma as a "respectable" venue—unlike protests—people who had never been willing to consider the reality of nuclear dangers had paid attention.

In our anxiety to give people as much perspective as possible on energy issues, CASE members were constantly on the lookout for interesting speakers. Barbara Geary was the one who found Bill Ward. When Bill was on the Kansas attorney general's staff in the 60s, he had worked with the attorney general to show that the salt beds at Lyons, Kansas, were unsuitable for a federally proposed nuclear dump. In the process, Bill had become a wind energy enthusiast, helping organize Great Plains Windustries of Topeka, Kansas. He hoped to enlist nominal consumer financial aid, and, eventually, federal grants to experiment with converting wind into electrical energy.

In appearances around the state, in June, 1975, he told of the current technologies that take advantage of solar energy—especially solar energy in the form of wind. "The heat of the sun causes convection currents which produce wind," he explained. "The wind churns a mill, behind which is a generator creating electric energy. Wind is cheap because nobody owns it—so far at least—and it causes almost no pollution."

"Don't tell me I can't!"

In 1975 the NRC was required to make records of their hearings on civilian nuclear power plants available to the public in a library in the area under consideration. (Now they are on microfiche.) Tulsa's downtown Central Library was selected as the depository for the Black Fox documents, and that year reports, information and reference materials started arriving. They were shelved as reference works in the business and technology section of the library, where circulating books on nuclear and alternative sources of energy were also available. As time went on, Black Fox occupied more and more shelves and numerous filing cabinets. Interested people started wading through these documents, and fellow CASE members and I spent so many hours in the

00 BALLOONS WILL E RELEASED, 200 OLUNTEERS NEEDED

A thousand balloons will be released Hiroshima Day, August 6, from Rocky Point near the proposed Black Fox nuclear plants near Inola to determine the possible path and extent of radioactive travel if the plants are built.

200 people are needed to hold the balloons while they are filled with helium and prepared for release. Cards will be attached to each balloon with the request that they be returned to Citizens Action for Safe Energy, Inc., when found.

Volunteers are requested to be at Rocky Point by 9 a.m., August 6, to help prepare the balloons.

_____ can be ____ eling 16 ___ Tulsa on ___ the sign: ___ Commo- ___ Recreation ___ ght at the ___ wo miles ___ ght again, ___ _en turn __

Black Fox Rally Set

rally against the ck Fox nuclear facility posed near In__ wil gin at noon, gust 7, at 3 verside Drive. Bring food tc are in a s ordinating ef lack Fox. Th peeches in c he nukes, a

Biologist sees genetic risks in power units

RADIATION in the environment in the latter part of this century has contributed to an increase in leukemia, an Oklahoma biologist argues in opposing nuclear plant construction. Dr. Connie Taylor, assistant prof __ ogy at Southeastern S __ and chairm __

Dr. Helen Caldicott

library searching for information that we sometimes felt we lived there.

As curious readers sorted through the documents, they started finding bits of information that contradicted what was being said in newspaper articles and PSO press releases. They started writing letters to the editor in an attempt to set the record straight. One such letter to the editor of the *Tulsa World* from a B. Cardwell expressed the natural bewilderment of any person who tends to believe what is in the papers: "My recent visit to the Tulsa Library revealed some very surprising statistics concerning the costs of nuclear power," it began.

The letter went on to mention its writer's discovery that 95% of the increases in construction costs for nuclear power plants in the past had been caused by delays in construction resulting from "late equipment delivery, poor labor productivity, labor shortages, equipment failure, and rescheduling of related facilities." Only the remaining 5% of the delays could be attributed to citizen action. In other words, power companies knew ahead of time that construction would actually be far more expensive than their published estimates.

The NRC documents also told us that they already knew about the dangers of the experimental and technological problems they were urging on us. Only 27 problems had been targeted for correction. They didn't know what to do about 183 problems, and they decided to ignore 44 more because of "manpower limitations" or because they didn't know enough about them. These were not just minor problems that could be considered negligible by a sensible person. A number of them were basic: Steam tubes that were likely to rupture, pressure vessels that could not withstand steam explosions, and turbines that were subject to malfunctions that could cause major nuclear accidents.

In November, 1974, the Monticello, Minnesota, reactor dumped about 50,000 gallons of radioactive waste water into the Mississippi River, some of which was sucked into the St. Paul drinking water intake before its gates were closed. That that kind of insult to public health was not unusual became apparent when we read the NRC documents: In just one fiscal year—the one ending June 30, 1974—the AEC found a total of 3,333 safety violations at the nuclear facilities it inspected. Who knows how many occurred at the facilities they did not inspect? Although many of the 1974 violations posed a threat to the public or to workers, the AEC had imposed punishments for only eight.

If the AEC/NRC itself considered these instances part of the normal risks of life, what, one could not help wondering, *would they consider unacceptable?* Day after day, we left the library more and more depressed and more and more determined to end the suicidal course on which greed and ambition had set our world.

Our shared knowledge brought us closer together, and we began to arrange our lives to allow for time to get together and bolster each other's spirits. We needed fun and laughter to counter all that gloom and doom. On Saturday, May 3, 1975, many of us gathered with our families at Rocky Point Recreation Area for a day of games and relaxation. At this beautiful park at the west end of the Black Fox site, we ate a picnic lunch and indulged in nonsense for a change. But it was not all nonsense.

Gary and Pat Laner of Claremore brought a tank of helium and a tent to the picnic. Nowadays, we know that deflated balloons swallowed by animals, often obstruct their digestive tracts and cause them to die of starvation. (Unlike the NRC, we have learned from our mistakes and no longer make that one.) But on that day, in our happy ignorance, we inflated and tied cards to balloons—filling a tent with them. When we were finished, we launched them all together into the intense blue of the windless sky—a breathtaking sight that still pierces my heart with remembered beauty.

The cards we tied to the balloons offered a dollar for every one returned and explained that we were investigating the distances and directions that radioactive effluents would travel from the plants. Responses came from as far away as Illinois and Kentucky.

Jesuit William Millerd, a physicist at the Washington, D.C., Center for Science in the Public Interest, took time out from his eleven-stop speaking tour of Oklahoma to spend that day with us. The next day he spoke at a conference room in the Central Library, decrying the "moral irresponsibility" of pushing nuclear energy to cope with statistically suspect energy-demand projections, when solar energy and zero energy growth were clearly feasible alternatives.

The more time we spent on the papers at the Central Library, the more CASE members believed that we had no choice but to intervene in Black Fox hearings. That summer our board formally authorized me to inform the NRC of our intention. I sent the letter to the NRC on August 5, 1975, certain that it would be weeks before someone wrote a formal response.

Imagine my surprise when just over a month later, on September 9, attorney Dow Davis telephoned to tell me that he and another NRC representative would be expecting to meet me in Tulsa that evening!

In retrospect, I think Jan Norris and Dow Davis must have hoped to catch us off-balance and demolish us immediately as a factor to be reckoned with. Fortunately, I was not alone. "Robert," I implored, "you have to go with me. I will need every weapon I can muster."

"You bet I will. I wouldn't miss this for anything," my shy warrior responded with emphasis, "but you'd better ask Ilene, too." Ilene called Ann Funnell, and they both dropped everything to drive over from Oklahoma City.

Jan Norris was the NRC Project manager for Site Safety and Environmental Analysis. I don't know whether he and Dow Davis believed what they were saying and just wanted to swat a gnat buzzing about their ears, or if they had begun to respect the power of ordinary citizens and dread the clouds of hornets CASE members were to become. Maybe they really thought they were compassionate advisors trying to save a misguided bunch of little old ladies in tennis shoes (and their blinded-by-affection husbands) from themselves. At any rate, they carefully explained that electricity from nuclear plants was good for the nation because it was very cheap.

Then they cautioned us that "atomic energy law does not allow an intervenor to prevent construction of a nuclear power plant." Emphasizing that the only power an intervenor has is to—perhaps—force the utility to change the site (if the original one is not suitable) or alter the design (if they could demonstrate design flaws), they concluded, "But if your goal is to stop Black Fox, you will be wasting your time, your energy and your money. You cannot stop a nuclear power plant."

What they didn't know is that the surest way to get me to do something is to tell me I can't—especially with something as important as stopping a nuclear power plant! No child raised by my mother and grandmother is willing to admit that he or she knows the word "can't." My mother's children tackle the job and finish it. Sometimes it takes good old American ingenuity to work out the problem, but we never give up.

At Critical Mass '74, I had learned that in most cases, utilities that built nuclear power plants eventually asked the utility commission for a rate increase to pay for the construction work they had already done and were intending to do. This money for "construction work in progress" was never mentioned when the utility companies were trying to convince people that they wanted a nuclear power plant. Utilities knew that if they told the whole story at the beginning, consumers would never stand for having to pay millions in increased rates for something that would not start operating for years and that they didn't even need.

"If there is no intervenor," I asked, "how many days does the hearing last?"

"About two."

"AEC law is not fair," I expostulated. "It is certainly un-American, and I believe it is unconstitutional. We will intervene, and we will stop Black Fox," I continued flatly.

"If we let you get away with this, PSO will be selling us electricity that is not only not cheap, but that will be the most expensive electricity we can buy. When you tell us nuclear power is the cheapest source of energy, you are not including the tax dollars for research and the subsidized insurance under the Price-Anderson Act. You are not including the costs of preventing sabotage and monitoring effluents for radiation or the costs of radiation-induced disease. And you are not even including the costs of decommissioning or of millennia of guarding nuclear wastes.

"We don't have the time to take the NRC to the Supreme Court," I concluded, "but we will use economics to stop Black Fox in the Oklahoma Corporation Commission hearing. We will intervene in the NRC hearings so we will be legally qualified to participate in the OCC hearing when that time arrives. There, we will prove that Black Fox electricity will be too expensive for people to buy. That is how we will stop Black Fox!"

In hindsight, it seems foolish of me to have told Norris and Davis how I intended to defeat them, but since they did not take me seriously, it didn't do any damage. After two hours of discussion they finally gave up on their efforts.

As we prepared to leave, I asked if there were any other steps we needed to take to become intervenors. "We'll let you know when the hearings will start," was their misleading response. Months went by before I began to realize the true horror of being an intervenor.

Meanwhile, Ralph Nader called together a second Critical Mass ('75) conference for November 16-18. This time, Ilene Younghein and Ann Funnell went with me. There, I finally talked face to face with a valued correspondent and advisor, Irene Dickinson; with Franklin Gage, she had founded the National Intervenors. Irene and Franklin had sent me loads of information, and I often consulted with them by telephone over pressing problems. And I met another new friend:

Australian pediatrician Helen Caldicott had come to Boston's Children's Hospital to work with children born with cystic fibrosis, the most common inherited childhood disease. Watching her patients die of respiratory failure and seeing children in other wards die of leukemia and cancer—diseases often caused by genetic damage—motivated her to speak publicly about her observations and to write a book, *Nuclear Madness*. She had also spoken out against the French atmospheric bomb tests in the South Pacific, arousing worldwide public indignation against them; finally, France announced that it would restrict its testing to underground sites. A tireless proponent of safe energy and a captivating speaker, Dr. Caldicott had put together a powerful slide presentation on the problems of nuclear power, that I was to use again and again in the years that followed. It was not until a couple of years after Critical Mass '75 that we actually managed to find the time to talk at length during her stopover at the Tulsa Airport.

Among the most useful sessions at the conference was a discussion led by Pennsylvania activist Judith Johnsrud on leverage points for citizen action against nuclear parks like the one proposed for Camp Gruber. She told us that one of the most important public health *dicta* was,

"When in doubt, DON'T!"

Another woman I met there, Jeannine Honicker, was, in 1978, to bring what I still consider one of the most important suits against the NRC. Few people realize that NRC rules do not protect citizens from radiation. They simply mandate that it be "as low as reasonably achievable," and some of those "reasonably achievable" levels are lethal. The Farm[1] provided legal and research assistance for *Honicker vs. Hendrie*,[2] in which Mrs. Honicker asked if it were constitutional to use standards that permit people to be killed. What, her suit demanded, was the difference between that and any other murder?

A constant undercurrent to all our discussions of safe energy and of tactics for defeating unsafe proposals was the reminder that just one year earlier, Karen Silkwood had paid the ultimate sacrifice to that modern god, Nuclear Energy. In her memory, we walked by candlelight from the Ellipse to the Capitol, where we listened to encouragement from Karen's father, Bill Silkwood, and were inspired by the quiet presence of her mother.

On December 23, 1975, the NRC began its review of PSO's request for a Limited Work Authorization permit. This would allow them to clear and grade the site, to install temporary support facilities (electricity, water, etc.), to make excavations, construct service facilities—and to install foundations for the structures they intended to build. *How dare they!* I thought. *By the time they finish all that stuff, they will have spent millions of dollars of the rate-payers' money, and everyone will think, "What a pity to let all that go to waste!"*

I mailed a letter of protest to the NRC, stating my objections to an LWA permit. I told them they should not allow any work to be done on the site, because CASE planned to prevent construction of Black Fox.

My reply was a letter from William H. Regan, Jr., telling me that public hearings would be held in the autumn of 1976, with a detailed description of the NRC licensing procedure, "Licensing of Nuclear Power Reactors."

"The laborer is worthy of his hire."

In 1976, shortly after the New Year, Robert and I were at the Claremore post office mailing some CASE letters when we were hailed by St. Cecilia's priest, Father Robert Pickett. "PSO has invited every minister in the greater Tulsa area to a luncheon meeting on January 14 at the First Presbyterian Church," he told us. "CASE ought to invite them to an earlier meeting and prime them with knowledge of nuclear power's dangers before PSO gets their ears."

I phoned Father William Skeehan at the Church of the Resurrection to ask if we could hold our meeting at his church on January 7. I composed an invitation, and Lavita Riggs and I addressed an envelope to every minister in the greater Tulsa

W Feb 4, 1976

Three Quit GE, Say N-Plants Too Dangerous

LOS ANGELES (UPI)—Three high-ranking General Electric Co. engineers who have designed and built nuclear reactors said Tuesday they resigned because the nuclear plants are too dangerous to mankind.

A General Electric spokesman said the resignations were a complete surprise. GE is one of the world's largest suppliers of nuclear power plants.

They were Dale Bridenbaugh, 44, manager of performance evaluation and imprivement, Richard B. Hubbard, 38, manager of quality assurance, and Gregory Minor, also 38, manager of advanced control and instrumentation. All live in San Jose with their families.

THEY SAID THEY WERE CON-

8 D   WEDNESDAY, FEB. 4, 1976

N-power plant protested

uclear experts (from left) Richard Hubbard, Gregory Minor and Bridenbaugh, who helped design and build nearly 100 nuclear plants for General Electric, have resigned from the company, nuclear power is too dangerous for further development. The aid they would campaign for passage of a statewide mora-on the construction of nuclear power plants. The three was engineering managers at GE's plant in S... photo)

area. Even though the roads that morning were icy, about 50 showed up.

We had put together a stellar roster of speakers—people who spoke well and had interesting things to say. Leon Ragsdale of the Ragsdale-Christenson Architectural Collective told them about the financial and technical feasibility of using solar energy to heat and cool church buildings. Radiologist Ralph Adams approached the nuclear debate as a moral issue, mentioning the $459 million in public funds that had been spent on nuclear fission research and development—much more than for any other energy source—and outlining the connection between that imbalance and our responsibility to safeguard the safety and health of this and future generations. Holding up the spectre of a nuclear-power police state, OSU professor Richard L. Cummins asked them to consider "Who is going to pay the cost and who is going to get the benefit?" and Ilene discussed the Camp Gruber energy park as a Trojan horse for the nuclear industry.

Mary and I visited with our guests as we served them the lunch we had prepared earlier in our nursing home kitchen. There I met the Reverend Marvin Cooke, a tall young man who ministered to a Methodist church on the far north side of Tulsa and who soon became a CASE board member.

At their meeting the following week, I heard PSO and the Atomic Industrial Forum tell the group that the required safeguards made the nuclear industry one of the safest in the world. Boldly, I had crashed the party, thinking I would soon be kicked out, but no one turned a hair. I sat there stunned by the untruths being foisted on the ministers, hoping we had given them what they needed to see through this disinformation.

PSO had brought in its big guns from all over the country. The first, E.R. Johnson of Virginia, complained about the American tradition of open forums, charging that we were still mired in debate, while other nations, unfettered by democracy, had made great strides. "I hope we can resume the leadership we once enjoyed," he concluded as I sat there thinking, *Yes, life would be much simpler for you if you could do exactly as you pleased without regard for the public good!*

The second, Dr. Roger Linnemann, president of Radiation Management Corporation of Philadelphia, admitted that plutonium is highly toxic but maintained that there was no evidence of higher incidences of cancer, leukemia or genetic defects in nuclear workers. (Almost twenty years later, the evidence is loud and clear that there are indeed higher incidences of these three killers among nuclear workers and their families!)[3]

Finally, Associate Professor D. Lynn Draper, Jr., director of the Nuclear Reactor Laboratory at the University of Texas, told the ministers that death by nuclear reactor accident is about as likely as being killed by a meteor.

All the speakers agreed with us that one great problem is spent nuclear fuel storage. They stoutly maintained, however, that the problem was "being worked out." (Almost twenty years later it has still not been "worked out!") It was left to Mr. Johnson to utter the most barefaced lie of the day: The total volume of waste produced, he told us, could be stored in a space 100 feet in length, width and depth.

When B. H. Morphis, assistant vice president and director of PSO's nuclear division, was asked about putting the cost of caring for the nuclear waste on future generations, he side-stepped neatly. "We do not want to deny future generations the natural gas, oil and coal that can be used for other things," was his answer. Mr. Johnson came to his rescue, noting that utilities were being asked to establish trust funds to provide for future surveillance and waste maintenance and ignoring the fact that the amounts in question were minuscule compared to actual costs.

Mr. Morphis tried to establish PSO's position on the side of the angels, speaking of its grants to OSU and the University of Tulsa for research in sun and wind power—as if the mere allocation of the money magically absolved PSO of any duty to explore implementation of the research findings.

The day was warmer and the roads less icy than a week earlier, and many more ministers ventured out that day. And they were given a catered luncheon. I felt terribly frustrated to be only one person in a small organization, trying to undo the harm a huge company was intent upon.

In Kansas there is a waterway called Wolf Creek. It flows into the Neosho River, which in turn flows into Oklahoma's beautiful Grand Lake, ultimately supporting recreation, fishing, and energy and water supplies for hundreds of thousands of people in Oklahoma. At the Critical Mass conferences I was told that a nuclear plant was to be built on the creek, and a number of the people I met there were trying to prevent its construction. I was especially worried about the plant, because a radiation spill into Wolf Creek would eventually affect many Oklahomans as well as Kansans.

Before 1976 I had attended a courtroom hearing only once in my life, and that was when I accompanied a nursing home patient to take care of some minor legal problem. I had no idea what to expect when the NRC hearings on Black Fox began, so when one of my friends told me there would be a hearing on Wolf Creek in February, I decided to go.

I boarded the plane in Tulsa wearing a light wrap in our balmy February weather. Kansas City greeted me with six inches of snow and a frigid, blustery wind! I had no time or money to buy anything heavier, so I just decided to hope the courtroom would be warmer.

The hearing was not as intimidating as I had feared. The intervenors had raised $2000 to make the thirty-five copies of every legal brief—drawn up by four young attorneys working *pro bono*—that the NRC required. Everyone who wanted to was given five minutes to speak—even I had my five minutes! I know the hearing was for a construction permit, but because I had no mental context to put it in, what I heard ran together into a blur in my mind.

That evening, however, was a different thing. Diane Tegtmeier offered the hospitality of her home to me as well as to a number of women from all across Kansas. Some of them I had met at Critical Mass; others I had corresponded with. But even those of us who had had no previous contact felt immediate bonds of support and understanding. There were many men fighting nuclear power, but I have often thought that the peculiarly feminine ability to form relationships without wasting time on deciding who is the pack leader—coupled with the fact that women seem not to be as worried about losing their jobs because of their political convictions—are the reasons that some of the most effective leaders in our struggle were women.

Diane Tegtmeier has degrees in physiology and chemistry. Quiet Mary Ellen Salava is a member of one of the pioneer Kansas farm families evicted by the plant. Margaret Bangs is one of the most determined people I have ever met: She stayed with the fight to the bitter end, and when it was lost (Wolf Creek went on line on September 3, 1985), she participated in the inevitable rate hearing. Kay Yoder and Jo Ann Klemmer were there from the Women's International League for Peace and Freedom, Kay carrying with her two suitcases of items she was selling to raise funds for the hearings; I wore one of her sun-and-windmill pendants as a talisman for the rest of the fight. Edith Lange and Wanda Christy were also there, as was Nancy Jack of the Sierra Club.

Civil engineer Ron Hendricks was an intervenor. Early on in my involvement in the battle against nuclear madness, Ron, a coordinator for Friends of the Earth and an expert on water issues, invited me to debate an official of their utility company at the university in Pittsburg, Kansas. Like most new converts, I thought I knew all the answers, but in the heat of debate, I floundered. Ron, who knew much more than I did, came to my rescue and saved the day!

It was sad to hear of the disruptions that Wolf Creek was causing to the lives of farm families and of the 10,500 acres of farm and grazing land that were being covered by a lake to provide cooling water for the monster. It was devastating, however, to discover the degree to which independent Kansans had been cowed into silence by fear of reprisals from the Wolf Creek supporters who controlled local jobs. The Midwest is often thought of as the last great bastion of democracy, but even it bowed to economic pressure.

A fragment of conversation I overheard at the end of the two-day hearing was for me, however, one of its most important episodes. I sat behind the four young attorneys as they discussed their continued participation in the hearings and as they worried about how they would support themselves if they continued. I could hear my mother quoting to me, "The laborer is worthy of his hire," and telling me that it was important not only to pay our hired hands but to feed them adequately.

I never learned how they worked out their dilemma, although I do know they continued to participate to the very end. What I did learn, was how important it was going to be to raise money to pay for attorney and witness fees for the Black Fox hearings.

Two days of Kansas City winter in a thin, knitted stole defeated my body, and I spent several days in the hospital with pneumonia. Even so, I felt—very prematurely—that things were beginning to look up.

By early 1976, the debate over the trade-off between the headlong rush to create electricity in ever-increasing quantities and the grave and often unknown risks involved had invaded the utility and NRC hierarchies. James Hooper, who had been nominated TVA president, told a senate committee, "It's a dead end with nuclear power, and we must take a breathing period." The AEC's original chairman, David Lilienthal, convinced that proliferation of nuclear technology was rushing us toward "impending disaster," mourned that, "I'm glad I'm not a young man, and I'm sorry for my grandchildren."

The nuclear industry was shaken on February 2, when three General Electric engineers, Dale Bridenbaugh, Greg Minor and Richard Hubbard, resigned. They charged that in its greed, GE was ignoring the safety of the civilian population and asked the NRC to review the country's fifty-six licensed nuclear power plants immediately. This was especially shocking because the three men, who had worked for GE all their professional lives (an average of 18 years) and who were giving up very lucrative positions in middle management, felt so strongly about the matter that they were willing to risk their incomes and professions to make themselves heard.

Bridenbaugh, Minor, and Hubbard were convinced that "the nuclear industry's rush to build more and more new plants" was an important factor in the near disaster at Browns Ferry in 1975. They predicted that the NRC's policy of exempting "previously approved nuclear plants from the safety requirements applicable to new nuclear plants" would inevitably result in other near misses and emphasized the very real possibility of a devastating disaster in the not-so-distant future.

One of the major features of nuclear safety systems is redundancy, that is, backup systems. The idea is that if one safety system fails another will save the day. However, if the redundant systems are placed too close together, a failure that knocks out one may knock out others. That is in fact what happened at Browns Ferry: A plant fire knocked out each of the backup systems at the same time as it shut down the first-line system. The three engineers explained that nearly every plant in operation suffered from the same fault, putting the whole country in jeopardy. In fact, they concluded, there is really no such thing as a 'safe' nuclear power plant. "We don't really know what is going on in a reactor," they explained, "and even if we did, there would still be human error and radiation-induced deterioration of materials to contend with."

In 1978, CASE hired these three men as our expert witnesses in the health and safety hearing before the NRC. They were invaluable in assisting our attorneys and in giving testimony before the Atomic Safety and Licensing Board.

A week following the resignation of the former GE Engineers, NRC Project Manager Robert Pollard resigned during a newscast on national television, provoking a media extravaganza that made at least a *pro forma* Joint Congressional Committee on Atomic Energy investigation inevitable. As was to have been expected, the committee concluded—in the face of all the available evidence that contradicted it—that nuclear power plants were perfectly safe. Pollard was NRC project manager at the Indian Point, New York, plant when he read an internal NRC document listing nearly 200 unsolved reactor safety problems. Despite the problems, the NRC was willing to let reactors continue to be built on the doorstep of New York City. Even in his shock and outrage, he was self-possessed enough to submit his resignation on live television news. He maintained to a national audience that he could not "in good conscience remain silent about the perils" of nuclear power plants. In an abrupt about-face, he offered his services to the Union of Concerned Scientists.

In another resignation, an AEC Idaho Falls reactor safety engineer, Carl J. Hocevar, left his job in a safety dispute with the AEC, then joined with the country's anti-nuclear movement.

Neither Honest Nor Straightforward

Early spring of 1976 was hectic at the nursing home. There was a flu epidemic, and for weeks I was so busy caring for patients that I had no time to read CASE mail. So when I finally had the time to sit down and read, the mail had been accumulating for several weeks. The impression Jan Norris and Dow Davis gave me in the fall of 1975 had been extremely misleading. I discovered with a sinking heart that a letter from the NRC with the procedures and time limit for becoming an intervenor had arrived weeks before. The filing deadline had come and gone!

"The law is made for man and not man for the law," I told myself and proceeded to demand, politely, that I be allowed to file a late request. I

Rick & Kathy Groshong

Citizen Group Enters Talks N-Plants

ea citizens group will in-
ture hearings involving
se to Public Service Co.
construct a nuclear
Inola.

ens Action for Safe
ng safety precau-
ental issues sur-
d nuclear plant

his week with
llatory Com-
one of its
who lives
interven-
it hear-
lary.

NF'

Tulsa Daily World June 4, 1976

DOCKET NUMBER
PROD. & UTIL. FAC. 50-556,557

DOCKETED
USNRC
MAR 26 1976
Office of the Secretary
Docketing & Service
Section

CITIZENS' ACTION for SAFE ENERGY, INC.
P. O. Box 924
Claremore, Oklahoma 74017

Secretary of the Commission
United States Nuclear Regulatory Commissi
Washington, D. C. 20555

Attention: Docketing and Service Sectio

Dear Sir:

This is to reaffirm our decision
on Public Service Company of Oklahoma'

Tom Dalton

later found that Ilene Younghein had filed independently without mentioning it to me, because she assumed that I had already taken care of the CASE application. After weeks of controversy CASE was accepted as an intervenor in Black Fox, as were Ilene and the Izaak Walton League.

The next problem was the list of contentions we were required to file. The term "contention" was a new one to me. I learned it meant "reasons I don't wish Black Fox to be built," "problems with the plant design," or "corrections needed in the site." Two women I met at Critical Mass, Kay Drey of University City, Missouri, and Mary Sinclair from Midland, Michigan, sent me the contentions they had used in their own interventions. Ilene submitted others drawn from the statements Bridenbaugh, Minor, and Hubbard made when they resigned a few weeks earlier. Although Tom Beam of the Izaak Walton League composed another list, financial problems later forced the league's attorney, a Mr. Blackwood, to withdraw their intervention.

In June, Common Cause issued a report questioning the objectivity of the ERDA and NRC in view of the fact that high percentages of their top policy makers had been recruited from the very companies that they were supposed to police. Concluding that there was a potential for serious conflicts of interest in both agencies, they urged the importance of making government agencies impartial and unbiased and kept at arm's length from the organizations over which they were supposed to keep watch. I added that to CASE's contentions.

On television one day, I saw Chicago attorney Myron Cherry appearing in dynamic opposition to the Palisades and Midland, Michigan, nuclear plants. When I telephoned him, he kindly suggested other contentions. He also made a point of cautioning me "not to assume that representatives of the utilities or Nuclear Regulatory Commission will be either honest or straightforward" with me. "Their interest, of course," he explained, "is to build a nuclear power plant, and you represent a barrier or a nuisance, depending upon how well organized you are."

Sudye Dalton offered her Tulsa home for a meeting to go over the contentions with NRC staff (including Dow Davis and Jan Norris) and PSO's attorneys. With me were the Reverend Marvin Cooke and a young attorney from Claremore who—I hoped—would become interested in representing us. I had never before been a part of such a conference and felt very inadequate. Davis and Norris peremptorily discarded a great number of our contentions.

My young attorney friend and Marvin Cooke were so upset by the incivility of the PSO and NRC attorneys that they left the room rather than be provoked into useless retaliation. They told me that that morning's experience had been all they could take. However, Mr. Cooke continued to work diligently with us to stop Black Fox. Later, Sudye and I agreed that the only person qualified to represent us was her then-husband Tom, an Oklahoma Sierra Club board member and a skillful environmental lawyer.

When in the fall of 1975 I had first asked Tom to represent CASE, he had said we would not be able to afford him, but after that conference I knew that somehow we had to find a way to change his mind. I asked Kathy Groshong (who, with her husband Rick, a geologist, co-chaired the local Sierra Club chapter) to approach him on our behalf. Although I did not know it at the time, Tom had also sent in a late request for the Sierra Club to intervene, but in mid-May he abandoned the Sierra Club initiative and accepted our request, bringing with him its full Oklahoma membership as loyal supporters of our cause!

If I had not overheard that conversation in Kansas City, it would have dampened my joy considerably when Tom gave me his cost estimate. He believed that for an effective intervention we would need at the minimum $10,000 on hand and an additional $10,000 later on. He told us he would have to employ experts in life sciences, utility and nuclear economics, and nuclear design and engineering with "experience in presenting views before courts or administrative boards" to consult with us on the petition and help us "prepare a written transcript of their testimony." His own fees were $35 an hour, $300 a day for trial appearances, and a retainer of $2000. In 1976, that was a fortune, but we knew he was worth it and that we had to find the money.

Tom warned us that "this will be a time-consuming and protracted effort, especially if an appeal is taken." It was, indeed, and his estimates were to turn out to be a tiny part of the actual cost!

Tom was not just competent, he was excited about this new challenge and put his whole heart into it, working long hours researching and then writing briefs. He was neither grasping nor mercenary. Knowing the hardship the expenses caused me and others and believing that the government should make it possible for ordinary citizens and not only wealthy corporations to be heard, he applied for financial assistance for CASE. The NRC—not wanting to encourage "a barrier or a nuisance"—denied his petition.

Tom's mind is both subtle and ingenious, but he is also thorough and quickly cuts to the heart of a matter. In my files is a copy of a June 14 letter he sent to OU geography professor and Sierra Club leader Marvin Baker. In this missive, he says that while he intends to pursue both environmental and safety issues contained in the fifteen-volume Black Fox Environmental Report and Safety Analysis, he feels that the strongest case that can be made, and the one most readily accepted by the general public, involves the simple need for power. Proving this case, he tells Baker, will take a thorough examination of everything from how need projections are calculated and the premises upon which they are made to the overall economics of energy industries.

Tom added one well thought-out, meticulously worded contention, admirable in its conciseness: "Intervenors contend that the Applicant and the Regulatory Staff have not adequately considered the relative costs and efficiencies of supplying power from solar or coal generating facilities."

Then he moved for dismissal of PSO's request on the grounds that the NRC lacked jurisdiction

1. because of the Safe Water Act of 1972, to authorize the discharge of radioactive effluent into the Verdigris River;
2. to proceed unless and until such time as the Environmental Protection Agency promulgates final standards and regulations which will govern the release of radioactive liquids or gases;
3. and, is constitutionally prohibited from making land use or land zoning decisions within the State of Oklahoma on non-Federal lands;
4. to license an activity which will violate state and local clean air and clean water standards, including non-degradation provisions;
5. to license Black Fox 1 and 2 unless and until such time as safe and adequate methods, techniques, sites, etc., are established for the disposal, storage, or other disposition of radioactive wastes.

The NRC, as was to be expected, denied the motion, but we were graduating from a nuisance to a barrier!

The End of Aunt Carrie's Nursing Home

Convinced that our cause was not only just but important to the future of people everywhere, we looked for every possible legal means of winning the day. One of those means was the initiative petition. Such petitions were being advanced in several states.

Jeffersonian democracy was alive and well in a number of territories when they became states. Distrusting the protections of representative democracy, these states had written into their constitutions a provision for ordinary citizens to initiate new laws. Thus, if the legislature became too arrogant or corrupt, the people would have some way of regaining control.

AUNT CARRIE'S NURSING HOME
A-1 CLASSIFICATION • 24 HOUR NURSING CARE
DOCTOR ON CALL
Furnished in Authentic Early American Pictures and Furniture
with matching Hi-Lo Hospital Beds and Chests —
Private Half Baths.
Physical Therapy, Personalized Beauty Care,
Laundry Service — Special Diets Available
Office: RO 3-4365 — Res.: RO 3-3494   Robert and
630 N. Dorothy, Claremore, Oklahoma   Dickerso

POST CARD

PLACE
STAMP

We wanted a law that would require public hearings and reports to be put before the state corporation commission before utility companies could be granted construction permits. That spring, Mike Males, son of Ilene's friend Ruth, returned to Oklahoma after working for some years in Oregon and was searching for just the right job. In the meantime, he wanted to do something for the future of Oklahoma, so he volunteered to do the necessary work of drawing up the petition. He spent several days in our home writing the "Utilities Siting Act," and CASE filed it on April 7, 1976.

Alerted to this attack on its new toy, PSO moved to counterattack. The proposed act would "ban inexpensive nuclear power in Oklahoma," their headlines screamed.

The Reverend Marvin Cooke countered that the proposed act didn't ban nuclear power but simply required that citizens be protected from potential damage and redressed for actual harm done. "PSO is complaining about the expense of adequate safety systems and insurance, and asking us to give up our right to protection and redress of damages," he explained, "leaving it to the public to draw the conclusion that, to PSO, profits were more important than people."

PSO riposted with a newsletter calling the proposed insurance requirements "unnecessarily stringent" and the safety testing requirements "exorbitantly expensive." "If the proposed utility act became law," they threatened, "it would shut down all work on Black Fox."

We were, of course, terrified by this frightful prospect!

CASE organized workers from our group and the Sierra Club to help gather the signatures needed to get the act on the ballot. We went to the League of Women Voters for expert help in signature-gathering techniques; Carol Oxley kindly offered us a series of excellent workshops. Robert had bought a van, thinking that we would start a child care center for Mary to

, TULSA, OKLAHOMA

Group asks hearings on N-plants

Oklahomans have a right to decide where and how utility power facilities are built in the state, a spokesman for Citizens Action for Safe Energy said today.

"Some 30 states have enacted laws that require public hearings and reports from state agencies before a power plant can be constructed," said Rev. Marvin Cook of Turley United Methodist Church.

operate and that the van would be used to transport the children. That summer our grandchildren were visiting from Massachusetts. Signe and Teddy Lemon were barely in their teens as was our Oklahoma grand-daughter, Sandra Snelling. Jim's children, J.J. and Melissa, were even younger. We took our grandchildren with us about the countryside. While Mary drove up and down the suburban streets of Broken Arrow and other small towns in Oklahoma, the older children split up in pairs to hand out flyers about our campaign against nuclear power and our initiative petition. The smallest child, J.J., went with me.

Patricia tells me that when Signe and Teddy talk about Oklahoma, it is that summer they remember. They knew they were doing something important. And it was one of the rare occasions we were able to spend any length of time together and really get to know each other during their childhood.

Although we worked diligently, we were not able to collect enough signatures to put the Siting Act on the ballot. However, we felt our efforts had been worthwhile, because the signature-gathering process had been another means of putting our side of the nuclear question before the public.

After the death of her mother, Delilah Hanes, Robert's youngest aunt, Sonora, continued to live in the magnificent old stone house her father had built in Sageeyah, the house where she had been born and where she had spent her whole life. Her nephews, who lived closer to the Verdigris River, had mowed the lawn and baled the hay for her, but she had pumped the water for the cattle and fed them for market. A girl-woman who never lost the wide-eyed innocence of her youth, she was—for generations of her young relatives—a ready source of sympathy and understanding. A week at her house was just the prescription for one who needed a respite from the pressures of adolescence.

Until her foster daughter, Mary Lee, grew up and left home, Aunt Sonora took care of herself reasonably well, but after that she often forgot to eat unless Aunt Ruth or one of the other relatives took her a meal. And even that was not enough. Unless they sat and ate with her, she would absentmindedly put the food down and remember it several days later when it had spoiled. She simply had no appetite.

All those years of poor nutrition took their toll on her mind, and eventually Aunt Ruth was appointed her conservator. The wonderful old house on the hill was sold. It went to a long-time friend of the Hanes family who valued it as much as anyone outside the family could—but it was, nonetheless, no longer hers. Aunt Sonora came to live in our nursing home. In her confusion, however, she was convinced that she had just gone down the road to visit a neighbor.

During most of her stay in the nursing home, the physical stamina and agility that had come from looking after her cattle made her a formidable walker. When she couldn't find her car, she would set out to walk home. But because her mind betrayed her, she could not identify landmarks. She just started walking—usually in a direction completely opposite to her goal. Once she got all the way to Lake Claremore, a two-mile walk, before someone noticed and took her back. Another time, she took a three-mile trek down Highway 20 to the Justus School. I do believe that if the sheriff hadn't called Robert, she would have wound up twenty miles down the road in Pryor. Robert was always the one who went after her, because she would recognize his Cherokee face, and he would be able to persuade her to go with him.

It broke our hearts to take her back to the nursing home. We had all loved her home ourselves, and we knew that the world must have become a confusing, terrifying place for her since she had had to leave it. We comforted ourselves with the fact that her sister Anna (Robert's mother) was there to visit with her and that Aunt Ruth came nearly every day. Did she still recognize them as her sisters, or was she searching for younger faces, the faces she had grown up with? We will never know, but we did know that there was no way she could go home. We were constantly worried that she would run off down the road and become completely lost or be hurt in an accident.

One day, Aunt Sonora went out the door and once more set out for Sageeyah. An aide quickly told Robert, and he set off after her. She was moving so quickly that he had to run down the street to catch up with her. I followed with some aides to help him bring her back, and thus I was there to see him collapse in the middle of the street.

For years, Robert had had problems with short-lived blackouts when he exerted himself too much, but he was a strong man, a man who had done hard, physical labor all his life on the farm. And when we worried, he would pooh-pooh our concern, growing angry if we tried to insist that he see a doctor about it. Eventually we would dismiss it from our minds. This time we could no longer dismiss it.

His doctors ran tests and told us he would have to have open-heart surgery. At that time, open-heart surgery was no longer uncommon, but it was not yet the everyday thing we think of now. We were appalled. Our worst fear was, of course, that he would not survive the surgery, but there were also a host of minor worries. What would happen to the nursing home? If we sold it, would we be betraying all the people who lived there, people Robert had known all his life and thought of as family? Would his own mother be properly taken care of? Would the buyer appreciate and preserve for the people there all the antique oak furniture I had bought at auctions in Nebraska and refinished for the nursing home? Who would look after the family concerns while Robert was unable to, or if—God forbid—he did not survive?

All those things would simply have to be taken care of. Robert was living on borrowed time. We started looking for a buyer we could trust to care about the people we had taken responsibility for. It took three months to find the person we thought would fill the bill. The papers were signed, but the money had not yet been paid when our borrowed time ran out:

One rainy day in the early spring of 1976, Robert paused in the nursing home garden to do a little weeding. Robert's father had grown up in a Missouri farm family, a family where vegetables were "garden sass" and flowers were fripperies beneath contempt. During their marriage, Mother Dickerson had managed to enjoy her one row of perennials—her two crepe myrtle bushes, her A.K. Viktoria and Cherokee roses, her Festiva Maxima peony, and her perennial sweet peas—only by dint of bodily preventing her husband from mowing them down. When they moved to the nursing home, things changed.

To be sure, my father-in-law babied his tomato and okra plants until they provided a substantial part of the diet of the nursing home, but half of the nursing home garden was given over to Robert's flowers. It was a riot of color—peonies, roses, zinnias, petunias, California poppies, and even hibiscus with their enormous, frivolous-looking blossoms. We had been careful not to laugh when we saw Robert's father cultivating them tenderly. But by that spring day, Robert had become their sole caretaker.

His mother stood leaning on her walker in the doorway, watching him as he did his labor of love, and she was the one who watched him collapse, this time with his face in the mud. Her vigilance was the only thing that saved his life. She was herself quite feeble by then, but love and terror lent her the strength to fly to her son's side and hold his face out of the mud until assistance arrived. Even then, Robert refused to have surgery. The most we could get him to agree to was to turn the nursing home over to the buyer without waiting for payment.

That was a difficult decision for both of us, because the nursing home had been our pride and joy. Mary and I had earned our Nursing Home Administrator licenses; we could have operated the home. As an RN, I could have continued to supervise the nursing duties.

However, I had discovered that few people were willing to spend as much time and energy as I did in the effort to stop Black Fox. Ilene Younghein lived in Oklahoma City, and she was working as hard as I in her way, but we needed someone in the Tulsa area who could work closely with our attorney. There were others who could operate the nursing home, but I firmly believed that no one else would stick with it to the bitter end of the fight to stop Black Fox. One cannot perform two big jobs and do either of them well. Robert and I agreed to give up the nursing home so I could devote all my time to the Black Fox fight.

On July 1, 1976, we turned over our nursing home to Duane Wood, and Aunt Carrie's Nursing Home became Wood Manor. At last I had the freedom to devote my full time to combating Black Fox. With Tom Dalton as our attorney, I felt we would succeed. For three years the profits from the nursing home (as well as calf sales from the farm) had been supporting my work with CASE. We did not receive payment for the sale of the nursing home until a year after the sale, and after the mortgage and all other debts had been paid, we had $200,000 left at interest in the bank.

CASE contributors seemed to believe that we were wealthy—and by the standards of the time, we would certainly have been well-off had that money stayed at interest. However, contributions never covered the expenses, and by the time Black Fox was canceled, the whole $200,000 had been lent or given to CASE for legal and expert-witness fees and duplicating, and I was left, like many others, with no income except social security.

Lawrence & Lottie Burrell

Black Fox delayed 3

TULSA (UPI) — An official of the federal Nuclear Regulatory Commission Tuesday said the first public hearing on the proposed Black Fox nuclear generating plant has been delayed for three months.

Dow Davis, counsel for the NRC, said the hearings, which had been s____ ____uled to begin in ____ ____ill not be held ____er.

Splitting the ATOM
lit
the
PEOPLE

Twenty

Delay, Delay, Delay

To people who have never been involved in a courtroom procedure or a congressional hearing, the process seems perfectly ridiculous. Instead of discussing things sensibly and deciding what to do, anything that has to go through a hearing turns out to be complicated to an unimaginable degree. Black Fox was no exception. As time went on, we became inured to delays and eventually came to welcome them on the grounds that every delay was just that much more time when Black Fox was not endangering us all. In July of 1976, however, I had not yet reached that state of serenity. No sooner, it seemed, had we had our informal discussion of contentions at Sudye's house than we had to go through all the annoyance—and, worse—the boredom of a formal hearing on them.

Three weeks after we signed over the nursing home, the NRC Atomic Safety and Licensing Board scheduled a special prehearing conference to decide who had a right to be involved, what could be discussed, and whether they should do anything about any of the motions that had been submitted.[1]

---

aring

ths

icant and other agreed upon a delay," Davis

tion was filed by

1869-1949

PEOPLE'S VOICE   Page 8-B   Sunday, August 20, 1978

Nuke Critic Defends Cred

litor's Note—Letters from lers are welcome, preferably ed, and should be no more than words. We reserve the right to contributions to meet space itations. Letters should be sent to Tulsa World, 318 S. Main St., sa, Okla. 74102.

rrie Responds

Vhen I started the fight against ck Fox five years ago I realized I uld be subject to criticism from sectors. I have tried to emphasize fact that we're fighting nulear ver, not Public Service Co. nor its ployees. I have tried to make it a ht of issues, not personalities.

an end to the proliferation of nuclear reactors. Scientists from all nations at a Conference on Science and World Affairs concluded:

"Owing to potentially grave and as yet unresolved problems related to waste management, diversion of fissionable material, and major radioactivity releases arising from accidents, natural disasters, sabotage, or acts of war, the wisdom of a commitment to nuclear power as a principal energy source for mankind must be seriously questioned at the present time."   Carrie Dickerson
Claremore

For Nuclear Plant

I thanked the Lord for the letters

are virtually non-polluting and not draw on irreplaceable fos fuels. There is solar, wind and wa power. Also power from orga wastes, such as methane fuel.

Any civilization that puts its fa in nuclear reactors would have to condemned as collectively insa One explosion or act of sabota would suffice to render large ar of the countryside permanen uninhabitable.

The U. S. Government spends lions of our dollars on nuclear re tors that use the deadly plutoni This same Government virtually nores the clean, safe solar ene that pours down onto the desert the Southwest free of charge, that could be harnessed by t

It was a bit like a school yard boxing match between a large boy who knows he can knock out his opponent with a blow, but is afraid that all the others will gang up on him as a bully if he does; instead, he tries to win the day with threats and flourishing. PSO's Michael Miller expressed the hope that the NRC would ignore anything we had to say ("proceedings would run continuously," were his words) so that they could get on with things with as little annoyance as possible ("proceed in an orderly fashion," was how he put it). He went on to say that he expected the large number of intervenors' contentions to be cut down substantially, or voluntarily withdrawn. ("Dry up and blow away," was what we knew he meant!)

Tom Dalton proceeded to frustrate his hopes. He explained that one of our contentions was that no meaningful discussion of the need for more electricity could take place unless the Draft Environmental Statement, which had just come off the press, was revised to separate the demands of wasteful, inefficient consumption from those of necessary consumption. He explained that if only necessary consumption were taken into consideration, the likelihood was that it would be discovered that we had more than enough energy production already. "As long as we continue to consider all electricity the same value, as long as we continue to ignore the end use of electricity, all we're going to do is continue to waste it. We will continue to create a condition of a throw-away society and an atmosphere in which it makes absolutely no difference whether it's efficient use or not," he concluded.

The unspoken part of his message, the part PSO, of course, heard correctly, was "Delay, delay, delay."

Another of our disputed contentions was that PSO had underestimated the price of the electricity to be produced by the plant, because they had omitted the larger portion of the high-level radioactive waste disposal costs. Tom argued that to disposition costs should be added those of millennia of surveillance, as well as the economic and social costs of cancer and other health problems and the resulting shortened spans of productive life.

It always puzzled me that PSO thought they could get away—in a public forum—with what amounted to telling the NRC what to do. At this hearing, Miller instructed the licensing board that he expected the environmental and site suitability hearing to take place in early 1977. Then the Limited Work Authorization could be obtained by the summer of 1977, and the construction permit issued in late 1977. Thus, the first reactor would go on-line in 1982.

Adjournment was an enormous relief to me—and, I suspect, to the people from PSO.

The day our hearing adjourned, a federal appeals court ruled on cases brought by the NRDC and the New England Coalition on Nuclear Pollution against Consumer Power Company's Midland, Michigan, reactor and Vermont Yankee's Vernon plant, instructing the NRC to strengthen its licensing rules for fuel reprocessing and disposal of nuclear wastes.

In response, the NRC announced on August 13, 1976, that no new full-power operating licenses, construction permits, or limited work authorizations would be issued until further environmental impact studies had been made! The Wolf Creek construction permit was delayed, along with five others, and nine full-power operating licenses were put on hold.

I was jubilant. At last, I thought, someone in a position to do something about it has been willing to look at the evidence. Anthony Roisman, the New England Coalition's attorney, spoke for us all when he said, "The NRC's actions are an admission that reactors are a serious environmental hazard." Few of us were, by that time, naive enough to believe that the NRC would not soon find some way of whitewashing the evidence, but every day that a plant was not on line was a day in which it was not producing radioactive waste.

Along with other CASE members, I continued to write frequently to our congressmen and to argue for alternate sources of energy when they or Oklahoma Governor David Boren made fact-finding visits to Tulsa. I had also written letters to editors of local newspapers from the beginning of my opposition to Black Fox. As time went on, however, the papers began ignoring my letters—whether because they felt people were tired of seeing the same name all the time or because PSO had put pressure on them, I have no way of knowing. Whatever the reason, I was not about to let my high profile get in the way of the message, so I started asking other people to sign my letters. In one batch of six letters I wrote, the five signed by other people were published, while the one that I signed was omitted!

One of those five letters referred to an article quoting the $1.5 billion cost estimate for Black Fox that PSO was still using in its publicity.[2] I quoted a paper I had found among those in the Tulsa Central Library, a PSO memorandum of understanding with Associated Electric Cooperative of Missouri, one of the Black Fox investors. The paper referred to "industry projections" as possibly "exceeding $800 per kilowatt (a total of $1.84 billion), depending on inflationary pressures on labor, materials, and components and the evolving standards and requirements" of licensing.

Despite this, PSO was still using $672 per kilowatt in the economic assessment it presented to the NRC—and the public. I came to believe that even though they continued to tout "inexpensive nuclear power," PSO really had no idea what the cost would be. I did not mention it in my letter, but even then I was beginning to wonder if the possibility that they had bitten off more than they could chew was beginning to occur to at least some people in PSO.

Certainly, they had realized nearly two years earlier, when they advertised for partners in Black Fox, that the cost was more than they could manage alone. Associated Electric became a partner in early 1975, contracting for 500 thousand kilowatts. However, PSO could not then admit that Oklahoma didn't need that electricity. PSO Vice President Martin Fate was given the job of announcing that the two units would be increased from the 950,000 kilowatts each, that Oklahoma supposedly needed, to 1,150,000 kilowatts apiece.

On May 14, 1976, Western Farmers Electric Cooperative of Anadarko, Oklahoma, also became a partner in Black Fox. Tom Dalton saw this as an opportunity to bring in Co-op customers in western Oklahoma as co-intervenors. This would not only broaden the base of opposition, but, since the filing deadline was November 29, would be another means of delay. The only people I could think of who might qualify were Lawrence and Lottie Burrell from Fairview, southwest of Enid.

When I was operating my bakery in the early sixties, I read that Lawrence Burrell operated an organic wheat farm and ground the wheat on a stone mill and that his wife, Lottie, made bread for their neighbors. I went out to Fairview to meet them and learned that we had mutual friends in Harry Krause and other members of our local Seventh Day Church of God. Over the years, the Burrells had become friends of mine, as well.

I kept putting off the trip to Fairview, hoping that Robert would be well enough to go with me. Finally, Tom had to remind me that we were running out of time, so one cold November day, I made a date to visit with the Burrells the next evening. Snow is a far greater hindrance in the South than in the North, because Southerners expect snowfalls to be rare and short-lived, so there is little provision for clearing the roads. The idea of driving through the six inches of snow on the road that day terrified me, but my nephew, Mike Cahalen, saved the day. He took me there in his four-wheel-drive vehicle and then set off on tour of the countryside while I visited.

When I got there, Mrs. Burrell went into the adjoining dining room where she could hear, but did not enter our discussion. I told Mr. Burrell how necessary it was for us to have another intervenor and why delay was our most important weapon. He accepted my reasoning but could not persuade himself that it was his duty to be on the firing line.

Finally, I told him that God would hold us accountable if we held our hands and permitted a facility to be built that would allow infants to be born with birth defects and cause suffering and death from diseases such as cancer and leukemia. After two hours of discussion we were both in tears—and Mr. Burrell signed the petition to become an intervenor!

Six More Months of Prehearing Conferences

Our delaying tactic worked. A second prehearing conference[1] had to be held to decide whether Mr. Burrell and four other petitioners would be accepted as intervenors, and what to do about some of the motions that had not been acted upon. Sherri Ellis, Wallace Byrd, and Ann Funnell had come forward when they heard through the Sierra Club or from Ilene that we were looking for additional intervenors from western Oklahoma; Clark Glymour petitioned independently.

We wanted as many petitioners as possible, because we had no idea what contentions the board would consider acceptable. But doing this was an expensive proposition for us, because every new intervenor had to have new contentions and new expert witnesses. Fortunately, we were responsible only for Mr. Burrell's intervention. We set about finding an expert witness.

Dr. Richard Webb, a nuclear reactor engineer, constitutional law authority, and former staff engineer for Admiral Hyman Rickover, believed the government had gone beyond its constitutional powers in promoting nuclear power. He had written a book, *The Accident Hazards of Nuclear Power Plants*, expressing his observations and convictions, and had testified for other intervenors, so we felt that he would be an ideal witness. He agreed to come from Massachusetts to attend the prehearing conference.

Dr. Webb had already spent four years on Admiral Rickover's Division of Naval Reactors staff when they were assigned by Congress to build the nation's first civilian nuclear power plant—Shippingport—outside Pittsburg. He had also worked at another plant in Michigan. "I thought the commercial application of nuclear power would be the answer to the energy shortage and air pollution problems," he explained. "Then I began to study how reactors can misbehave, inquiring into the accident potential of present-day reactors such as those that would go into the Black Fox plant. I concluded that we don't understand reactor explosion hazards."

Dr. Webb never traveled by air, so we paid for his round-trip train ticket to Perry, Oklahoma, the nearest train station that connected with Massachusetts. It had snowed earlier in the week, and the roads were packed with ice. Robert knew our car would not make that hundred-mile trip under those road conditions, so he loaded every concrete block we had into the back of the old farm pickup to provide enough weight for reasonable traction. We set out, in what we thought was plenty of time, to meet the train at Perry.

The roads were even worse than we had expected, and by the time we arrived at the station, the train had come and gone. We searched in vain for Dr. Webb and finally went to a nearby restaurant, thinking he might have gone there. There was no Dr. Webb, but we took the time to eat breakfast and go to a lumber yard for more concrete blocks before we went back to the train station for one more search. This time we left not the most obscure nook unsearched. Dr. Webb had been there all along, sitting in a secluded area behind the station, out of public view.

The heater in our old pickup was long past repairing, so we bundled up

Dr. Wallace Byrd

6 SECTION B

N-Plant Hearing Delay See

More delays are likely in the hearing schedule for Public Service Co.'s proposed Black Fox nuclear generating station near Inola.

PSC, in seeking a construction permit for the Black Fox project, had hoped that hearings on Black Fox's environmental aspects could begin this month. But an official of the U.S. Nuclear Regulatory Commission said Tuesday hearings may not come for

"The problem has been the availability of information on water supply for the Black Fox plant," said Dow Davis, of the NRC's legal staff in Washington.

DAVIS SAID THE NUCLEAR plant water issue has been the main reason for the NRC delay in completing its final environmental statement on Black Fox.

He said the latest estimate is that the environmental impact statement on the project will be out by the end of this

Safety and Licensing Board will conduct the session.

Bert Morphis of PSC's nuclear sion said any delay expected in hearings is "minor" and will not a the nuclear plant construction dule.

Morphis explained that it is no lack of a contract for water for Fox that has held up the NRC environmental statement.

"It was simply whether we show that water is physically a to meet Black Fox requirement said.

in quilts we had brought along to keep ourselves warm, and started the long trek back to Tulsa. We kept our fingers crossed, hoping we would reach our destination before adjournment. The road was so icy that I feared at any moment we would land in a ditch or be pitched over a precipice. I had no emotional energy left over to deal with the terror that Dr. Webb expressed in incessant chatter.

Robert was so tense that at one point, when he almost lost control of the pickup, he so forgot himself as to all but shout, "Please shut up!" Thanks to his skillful driving, however, we reached the courthouse in Tulsa in one piece—and before noon. That was the most terrifying trip I have ever experienced. Robert was exhausted, and Dr. Webb was limp as a dishrag with fright.

Our bodies were so frozen we could hardly walk. I made it as far as the ladies' room, where I discovered Ilene with Ann Funnell and Sherri Ellis. Because of the slick roads, they had only just arrived from Oklahoma City. When they walked in, it had been during a recess, and Bob Mycue, a *Tulsa World* reporter, had told them the hearing was almost over and it would be best for them to not go into the hearing room. He told them that if they did not show up the proceedings could not continue, which would create one more delay. They decided to stay in the ladies' room to keep the NRC board from learning of their late arrival. Undaunted, I continued to the hearing room—only to be met with a PSO attorney distributing their motion for a Limited Work Authorization.

That motion struck terror to my heart. An LWA permits site preparation work (clearing trees, moving dirt and so forth), before a construction permit is granted. Once work is begun on a nuclear power plant, it gains a momentum that is nearly impossible to reverse. Fortunately nothing could be done about the motion right away, because National Environmental Policy Act issues had to be considered before an LWA could be issued.

By the time I arrived, Tom Dalton had already brought up the fact that the Cherokee Nation owned the river beds in the area and would have to agree before PSO could use the river water the City of Tulsa was preparing to sell them from Lake Oologah. Tulsa was also proposing to sell sewage effluent to PSO. Tom questioned whether PSO would be able to discharge the sewage effluent into Bird Creek without extraordinary—and expensive—treatment to meet Clean Water Act guidelines.

At about 6 o'clock that morning, Dr. Wallace Byrd had left Coalgate, in southern Oklahoma, with his counsel John Axton. Even so, they arrived late. The Burrells, too, had risked their lives traveling 200 miles, half of them over the same hazardous roads we had traversed, but they had arrived on time.

Hearing Chairman Daniel Head agreed that the change in Black Fox ownership justified proper consideration of the petitions for intervention. But he decided to wait until February to discuss the new contentions, because the board had not sent out some papers properly and some petitioners were absent—after all those hair-raising journeys! We comforted ourselves with the thought that while it meant another long trip and still more expense, it also meant still more delay as well—a delay that might not have occurred had Ilene and her group not remained in the ladies' room.

Tom Dalton suggested that the board question Dr. Webb about Mr. Burrell's contentions, since the financial burden and scheduling difficulties might make it impossible for him to return in February. Chairman Head refused and dismissed the hearing at 12:10—but only after Miller had sneered at Webb, implying that he had no credentials and had not even written his own testimony. Dr. Webb was so furious at having come all this way, placing his life in peril and wasting his time and our money all for nothing, that he could not contain himself—he told the hearing officers just how he felt. In no uncertain terms!

To Robert's and my enormous relief, the Burrells kindly offered to take Dr. Webb with them in their warm, comfortable car to the station in Perry. Their route was still precarious, and the people from Coalgate and Oklahoma City also traveled treacherous roads, but everyone arrived home safely, including Dr. Webb.

As we had expected, Dr. Webb was unable to attend the February 15 prehearing conference, but this time all five of the petitioners were there with counsel. Tom Dalton brought James Kouri to help represent CASE and Mr. Burrell.

The new intervenors were given an opportunity to explain first why they were qualified to intervene—that is, how they would be affected by the plant. Then they went on to explain which contentions they wanted considered and what technical expertise they could bring to bear on them. I already knew some of the new intervenors. Ann Funnell, for instance, had been at the meeting in which Dow Davis and Jan Norris tried to frighten us out of intervening.

As a Western Farmers Electric Cooperative consumer, who also attended church-related activities near Inola, Mr. Burrell was clearly qualified. However, the board eventually rejected two of his contentions that had been suggested by Dr. Webb on the grounds that the government was responsible for appropriate waste disposal and that his worry about a severe (Class IX) accident was really a challenge to the design criteria. (Design criteria were not allowable subjects of discussion at the hearing.) Mr. Burrell's contention addressing Anticipated Transients Without Scram *(q.v.)* was eventually accepted, and he became an official intervenor.

PSO attorney Miller tried to convince the board that Dr. Webb was not qualified to be an expert witness and complained that accepting new intervenors would delay the actual hearing, originally scheduled to begin on May 23, 1977.

A shocking professional experience had prompted Dr. Byrd, a member of the Oklahoma State Board of Health, to intervene. One of his patients had lived in Nevada downwind of an atom bomb test and had died of radiation poisoning. The patient's water supply had been a rooftop cistern in which radioactive fallout had settled. Watching that patient die had sensitized Dr. Byrd to the dangers of imperfectly contained alpha emitters like plutonium, that could be produced in the reactor. His expert witness, Dr. Edward Martell from the National Center for Atmospheric Research in Boulder, Colorado, was an expert on alpha-emitter aerosols.

The NRC had already decided that producing plutonium in a reactor was acceptable, alpha-emitter or not, so Dr. Byrd's petition to intervene was denied.

Professor Glymour, whom I never met before or after the hearing, told the board that his interest was straightforward economics. As a

Western Farmers rate payer and a chemist, he was worried that safety defects would cause frequent shutdowns and damage the cooperative's economic structure. He had been prompted to intervene by Dr. Webb's book and Mike Males' economic analysis of Black Fox.

Roberta Ann Paris Funnell of Oklahoma City, a medical photographer, is a descendent of one of the oldest families in Claremore. Her great-grandfather had owned and operated a stone quarry. Its profits had afforded him the ownership of an entire city block in Claremore. The board questioned her at length in deciding whether to admit any of her contentions. Among the subjects of those contentions were Class IX accidents, the unsafe conditions of the railroad tracks over which the radioactive wastes would be transferred, the environmental effects of waste handling, and the degradation of the environment by chemical wastes like sulfuric acid that would be dispersed from the cooling towers into the river.

Ann's turn was to come up right after our lunch break. During the break, PSO's Miller told Ann that he hoped she didn't have much to say, because he had to catch a plane to get out of there and didn't wish to be delayed.

I wasn't aware of that conversation, and was surprised that Ann read into the record everything she could think of, things that really didn't pertain to the points she would have been expected to discuss. Then she made an issue of PSO's contempt for the rights of the public, reading into the record a PSO instruction to the board that "any late intervenors should be compelled to take the schedule as it exists," underscoring PSO's impatience with any viewpoint other than their own.

When the board asked his opinion, Tom Dalton argued that the Constitution "does recognize higher values than delay and mere administrative ease" and supported her intervention. The board nonetheless rejected her as an intervenor.

Sherri Ellis, whose family owned land near Guthrie serviced by Western Farmers, was Karen Silkwood's roommate when their apartment—and their bodies—were contaminated with plutonium. She, too, had worked in the Kerr-McGee Nuclear Facility at Crescent. Her experiences there made her unable to believe in the adequacy of monitoring stations, employee safety measures, or plant security at Black Fox. Some of her contentions duplicated previous ones, and her intervention, too, was rejected.

Once more, PSO attorney Miller instructed the board, telling them that PSO expected findings by the end of the May 23 hearing that would entitle them to a Limited Work Authorization.

The day ended with a spirited exchange between Miller and Tom Dalton. In an effort to delay, if not prevent the LWA from being issued, Tom questioned PSO's compliance with water quality standards. Miller insisted there were no water standards with which PSO would not comply. Tom responded that our expert witness would show that, without question, PSO would be out of compliance. While Tom did not win his point and Chairman Head eventually granted the hearing motion, it was delayed for another three months, because another prehearing conference was called on June 27, a month after the originally scheduled hearing date!

That conference, presided over by a new chairman, Mr. Wolfe, began with a minor skirmish: Miller reported smugly that PSO's negotiations with the City of Tulsa for water were "at a very advanced stage" and promised to keep the parties informed. Tom resorted to seemingly desperate measures, asking if there were "any sense in going forward with the hearing" while PSO had neither fuel nor water. Miller protested and was upheld by Chairman Wolfe.

Then they got down to the business of arguing about our contentions. They started with PSO's selection of a site so near a large population center—only 23 miles from downtown Tulsa and approximately 13 miles from its then-current boundary—and ended with erosion and herbicides. Miller, of course, argued that Inola was a sparsely settled community. He was countered by Tom saying that, "Tulsa doesn't even know where its population boundary area is" now, let alone where it will be by the time the plant could be built. In their exchange, they did not even consider Claremore, Catoosa, and Broken Arrow, all exurbs of Tulsa with rapidly expanding populations and all less than 13 miles from Inola. I felt like asking, "Don't we count?"

PSO criticized CASE for not having expert witnesses to address all of our contentions, and Tom countered that we were "being penalized for being poor." PSO attorney Murphy, a young man who had a disconcerting tendency to dance about the courtroom like a boxer, criticized the expert witnesses we did have on the basis of religion and supposed emotional instability. (One of our witnesses, Dr. Rosalie Bertell, was a Catholic nun, and Dr. Webb had been goaded into an outburst in the previous hearing.) *Ad hominem* arguments are, of course, the last resort of those who know that their position is logically untenable, and Murphy undoubtedly knew that our witnesses' intellectual credentials were unassailable.

The flourishing was over, and on August 22, 1977, the hearing was to begin. We had bought three months of delay!

Twenty Two

The Sun in Splendor

Like an ever-changing kaleidoscope, CASE activities and our family's activities were continually intertwining and rearranging themselves. We held monthly CASE meetings—sometimes in members' homes, sometimes in libraries—while between times we continued to conduct public meetings where we showed our films and did our best to educate the public about the problems of nuclear power.

By June of 1977, Robert's heart could no longer pump the blood forcefully enough through his body, and his lungs were filling with fluid. He knew he was dying, and finally, reluctantly, he agreed to allow the doctors to try to save him. His mother was by then so frail that we feared for her life if she knew the danger he was in, so we told her that he was away on business. In his absence, she became more and more fretful.

A few days before Robert went into the hospital he wrote a $3000 check to our attorney for CASE fees. We had still not been paid for the nursing home, and money was getting tight. Handing the check to me, he said, "Carrie, we're almost bankrupt. You're going to have to find another way to finance the battle."

In my prayers that night, I asked, "God, do I have a talent that can help me raise money?" Leaving it in His hands, I went to sleep, and at five o'clock the next morning awoke from a technicolor dream of a beautiful sunburst quilt. My body wanted to turn over and go back to sleep, but my heart said sternly, God has shown you the answer to your prayer. If you go to sleep the memory of the dream quilt will vanish. Get up and sketch it.

I used the old dishpan to outline the sun on butcher paper. A yardstick and paper folds delineated the rays. It was long past dawn by the time I had finally recorded the technicolor dream quilt to my satisfaction.

After the first flush of gratitude had passed, doubt crept in. "Did you really mean it, Lord," I asked. "I haven't done a stitch of quilting since I left for college forty-two years ago. I will have to rely on You to help me remember Grandmother Perry's teachings—her tiny, tiny stitches, the way she used her thimble."

Putting my trust in the Lord, I went on as if I had been quilting every day of my life. Mary and I searched the fabric shops for just the right shades of yellow, red, and orange to turn my dream into reality, and through the long hours at Robert's bedside, I cut pieces and started trial constructions. Robert spent three weeks in the hospital before his lungs were clear enough for him to survive surgery.

I called our daughter Patricia at work in Massachusetts and asked her to fly to Oklahoma so that she could be sure of seeing her father for one last time. Patty's arrival distracted her grandmother's mind somewhat from Robert's absence, but Mother Dickerson could not understand

why—with his daughter in Oklahoma—Robert hadn't cut short whatever it was he was doing and come home.

On the morning of Robert's surgery, our four children were gathered with me in the waiting room. Throughout that endless day, our hearts were filled with prayers that God would guide the surgeon's hand. Our own hands were filled with the colors of life and joy, and we knew that our prayers would be answered. They were. Robert's surgery was successful! His life was given back to him—and us—with a new, artificial aortic valve. In that time of joy and relief, Patty christened my dream quilt pattern the *Sun in Splendor.*

When Patty and I were allowed to visit Robert in the intensive care unit of the hospital, he looked like death, but we knew the worst was over when, unable to speak or smile because of all the tubes in his nose and throat, one eye closed in a slow, deliberate wink. We both burst into relieved laughter. Robert was using the only language he could to reassure us he had just escaped his own deathbed!

Our joy was diluted when Robert's surgeon reported what he had found during the surgery. Robert's mother had told us that he had been a sickly child. That must have been because he had been attacked by rheumatic fever bacteria when he was small. Nowadays, antibiotics have tamed that monster, but in the second decade of this century it was still free to leave calcium deposits on Robert's heart valve and within his heart muscle, and through the years it had continued its lethal work. The surgeon told Robert that the calcium deposits in his heart muscle would cause a stroke within four years.

By then, not even Patty could calm Mother Dickerson's anxiety about her son, so we told her that he had had the surgery and that it had been successful. We could see the tension leave her. By small degrees, confidence that he would return to her suffused her with new hope. Had she been able, we would have taken her to the hospital to visit him, but her condition prevented her from riding in the car.

I worked on the quilt wherever I went. When I was a child, it had seemed to take forever to sew together the tiny pieces of fabric and even longer to quilt the completed assembly of top, cotton batting, and lining. The reality was just as I had remembered!

At that time, a museum-quality quilt might have sold for a thousand dollars or more, but that would have been only a drop in the bucket compared to the amount of money we would need to raise. I knew there was far more potential in using it as the prize in a drawing. However, Tom Dalton warned me that I could be prosecuted for holding a raffle. "Which is worse," I demanded, "going to jail because I owe you money, or going to jail because I've raised the money to pay you?"

I had no desire to go to jail for either reason, so I called the Tulsa District Attorney and laid my dilemma before him. He assured me that the law was usually ignored if the raffle was for a good cause. "But," he cautioned, "if we receive a complaint, then we'll have to prosecute."

My qualms disappeared. What better publicity could there possibly be than for me to go to jail because PSO turned me in?

Every moment I had with Robert seemed so precious that I decided to leave it to others to attend the fourth prehearing conference on June 27—less than a week after Robert's surgery—and the ACRS hearing three days later. Robert would have none of it.

"You can't miss those hearings, Carrie. They're too important. I'll be fine," he insisted.

So, while I sat in the conference room collating papers and quilting, Susie Birkhead sat in the hospital room with Robert, talking, reading to him, adjusting his pillows, or just watching while he slept. During the ACRS hearing, Joyce Nipper performed the same functions. They were not the only CASE members who expressed their affection for him so concretely. The number of those who helped was limited only by the importance of his being able to rest quietly.

Jo Ann Kelly and Esther Belz lived only a few blocks from Hillcrest Hospital, and when I needed rest, I went to their home for an hour or two of sleep before going back to Robert's bedside. How wonderful to have friends we could count on when we needed them!

When Robert was finally able to return home, his first stop was to visit his mother in the nursing home. As she clasped his hand in hers, her face might have been that of a madonna in a *pietà*. He had returned to her and she was content.

The hearing, which was with a subcommittee of the NRC Advisory

Committee on Reactor Safeguards, was held in a meeting room at the Tulsa Holiday Inn. Outwardly, it could have been any of the dozens of meetings of various professional associations I had attended. And the committee members, Chairman Max Carbon, J. Carson Mar, and Chester P. Siess, seemed to believe it was no different. While for Vaughn Conrad, John West, Charles Crane and Martin Fate of PSO and the Black & Veatch and GE representatives there was a considerable economic stake, their demeanor said, "It's all in a day's work." Tom Dalton, too, maintained a façade of easy composure, one that I strove to emulate as we sat silently through the discussions.

Each of the parties opened with an introduction of the organizations they represented and their experiences and investments in nuclear energy. PSO's assets included a nuclear division comprising about fifty people. Black & Veatch, a consulting engineering group from Kansas City, had never before had total responsibility for the engineering design of a nuclear power plant, but they had twenty people working on it. General Electric was, of course, one of the principal suppliers of nuclear reactors in the country.

Then the gentlemen from the ACRS committee listed the information and discussions they wanted to hear:

- What would happen if an earthquake cracked open the container holding the coolant that prevented the reactor from getting so hot that it melted?
- What special procedures would be necessary to produce a durable structure on the hard-to-work-with sandstone at the site that was so easily damaged by weathering?
- What would happen if there were a tornado or if a dam broke upstream and caused a flood—two very real possibilities?
- What provisions could be made for emergencies ranging from ordinary accidents to large releases of radioactivity?

For five and a half interminable hours, Tom Dalton and I sat and listened to all the things that could go wrong at Black Fox, all the things that could cause an "emergency event"—and to the frightful ramifications of such an event—couched in terms that made them sound no worse than breaking eggs or dropping a jug of milk. The thought that kept returning to my mind was, *How much cheaper, how much easier, how much safer, if they simply put up windmills instead of spending all this time, effort and money figuring out how to cope with a nuclear disaster.*

Throughout their insane statements, delivered with all the solemnity of biblical pronouncements, my fingers were busy creating the *Sun in Splendor.* Only the repetitious motion of the needle, weaving in and out of the brilliant fabrics, kept me calm. As they listened to each other droning on throughout the day, people's eyes would survey the room searching for some relief from boredom, and they would see that my fingers were occupied. Whenever there was a recess, they would come over to see what I was doing and why. I always explained that it was a fund raising project that also served the purpose of helping me keep my sanity in the face of what I was hearing. They would exclaim about how beautiful it was—and some even took raffle tickets!

Martha and Mary

The dancing and flourishing of metaphorical fists occupied all sides of the debate for the next two months. PSO's opening salvo on July 7 was a declaration in the *Tulsa World* of their position that Black Fox could be built and operated without undue risk to public health and safety.

Three weeks later[1] I had my say in the same newspaper: "We aren't naive enough to believe the NRC will be very receptive to what we have to say. Our best hope lies in alerting the public to nuclear plant dangers by making a record at the hearing. We know that just getting up, waving banners and shouting is not going to stop the Black Fox plant." To its credit, the *World* published a fair summary of our objections.

One of the major advantages PSO had with the public throughout the fight is that people like to feel progressive, modern, and in control of things. This has been a feature of the westward movement across America since the days of the pioneers. To have the courage to leave everything known and strike out into the unknown— and also persuade others to do so— often takes a sense of being part of something new, extraordinary and worthwhile.

This has been compounded since the sixties by a kind of societal Thumperism; Thumper was the rabbit in Walt Disney's *Bambi* who admonished, "If you can't say anything nice about someone, don't say anything at all." The opponents of Black Fox were, therefore, easy targets for being portrayed as naysayers—antediluvian dinosaurs, who were not

Sister Testifie[s] On Black Fox

Dr. Rosalie Bertell is a member of her hon[...] the Grey Nuns of the Sacred Heart Memorial and a cancer research scientist. She York, not came to Tulsa on her vacation from [...]ahoma su[...]

Dr. Rosalie Bertell

only stuck in the Dark Ages themselves but who couldn't bear the idea of their neighbors getting ahead. Unfortunately, this mindset ignores the fact that if we wait to scream until the wolf has us by the throat, we will be dead.

PSO chose the Sunday paper the day before the hearing convened for its next contribution: "PSC Reaches Overseas for Huge Reactor Part." The article featured a depiction of a reactor pressure vessel shell—twenty feet in diameter and the height of a six-story building—resembling the ones under construction for Black Fox in Rotterdam. The first was scheduled to leave Holland in 1980 for Memphis, where GE would install reactor components, turning it into a giant tea kettle to produce steam to turn turbines and produce electricity.

An important accompaniment to the public opinion battle that summer of 1977 was a crisis in the

Hearing
ear Station

well Park frightening
ffalo, New for life
: in the Ok-
stify a
Nuc
I.

rience.
itted te
ow leve
the U.S.
mmittee
and be-
ly Con-
ion cal-
he U.S.
atives
many is
a citi- hol
ved in to a
leve
pres- accep
s are "I c
a to- using
nts. mechar

Carrie with Ilene Younghein

women stand against PSO hearings on nuclear plant

Tulsa water system. On July 25, Tulsa Water Commissioner John Thomas renewed an earlier plea for voluntary water conservation. Tulsans had been using over 100 million gallons of water a day for more than four weeks, a volume greater than the system's normal operating capacity and one that could not be maintained indefinitely.

PSO had been negotiating with the City of Tulsa for three years to buy cooling water for Black Fox and the coal-fired power plants they were building at Oologah. The Friday before the hearing, the Tulsa City Commissioners once more delayed making a decision on the contract. Finance Commissioner Norma Eagleton still had not found satisfactory answers to questions she had posed months earlier.

According to the contract, water for Black Fox would have come from the city's share of the Oologah Reservoir (on the Verdigris River in northwestern Rogers County) and from Tulsa's treated sewage. The idea of selling sewage effluent elicited a lot of excitement: "Where else can one dispose of one's sewage and get paid for it?" demanded a bedazzled Tulsa TV anchor.

At the same time, the Tulsa League of Women Voters, also represented by Tom Dalton, was suing in federal court to prevent the Corps of Engineers from contracting for water storage rights at Oologah until they had an environmental impact statement on the sale of Oologah water to PSO.

PSO went into the Environmental and Site Suitability[2] hearing with the complete confidence and expectation that it would end with the issuance of a limited work authorization and the beginning of site work. Given that the NRC had shown itself to be pretty much in the nuclear industry's pocket, their confidence was reasonable—and ultimately justified. However, the NRC people we dealt with were, I am convinced, neither venal nor uncaring.

In the early years of the Atomic Age, our research physicians—very like many of those in charge of Hitler's concentration camps—told themselves that they were advancing the cause of science, and thus of human progress, when they injected healthy people with radioactive substances that eventually killed them. Because biophysics was in its infancy, they had the further excuse of not actually being sure that their victims would be harmed. In the light of what we now know, they were monsters of inhumanity. In their own minds, they must have been bold explorers.

I have no idea what rationalizations the NRC—or PSO—officials had used to build protective walls around their consciences. I do know, however, that they were savvy enough politically to realize that they could not with impunity give the appearance of riding roughshod over citizens' rights.

To give ourselves any chance at all of success, we had to hold their attention for as long as possible. We were willing to make them angry, frustrated, and bored, if that was what it took. What we weren't willing to do was to let them win by default. Those NRC hearings which—had there been no intervenor—would have been expected to take less than a week eventually lasted more than eighteen months. The delays were a double-edged sword. They cost us an enormous amount of money, but they ultimately made it possible for us to win.

By now I should have found the atmosphere of the hearing room as comfortable as an old shoe, but somehow this was different. The three board members were seated at a table on the stage. Three rows of tables were set up in a large area below the stage and perpendicular to the board members' table. Tom Dalton sat unaccompanied at the end of the center row of tables, facing the three board members. The same sort of shiver that went down my spine must have attacked him as well: "I feel quite alone," Tom told the board members, pointing to a battery of half a dozen PSO attorneys and technicians on his left and a similar battery of NRC lawyers and staffers on his right.

His isolation was very brief. During the hearing our expert witnesses sat at the table with him, and often Ilene Younghein acted as his assistant. She would pass Tom handwritten suggestions for questions to ask the witnesses or mention to him points that occurred to her as she listened to testimony. Ilene took time to read everything that had any bearing on the hearing, and she had the kind of mind that let her talk clearly with Tom and the witnesses about the smallest details. Without her, we would have been lost.

Ilene was the "Mary" of our team, and I was the "Martha." My hands were never idle. Much of my time was spent at a table in the back of the

room, collating and stapling copies of our expert witnesses' testimonies that
I had duplicated earlier, often having stood at a copy machine far into the
night. At least thirty-five copies were required of each document and sometimes more. Any free time I had was spent working on the sunburst quilt,
and during recesses I took contributions for the quilt drawing.

Keeping up with fund raising, duplicating testimonies of our expert
witnesses and our attorney's briefs, carting expert witnesses from and to
the airport and their lodgings, and looking after my recuperating husband kept me hopping. At the end of a long day filled with many different projects, I often felt like a basket case. If Mary had not been there to
relieve me, I am not sure how I would have survived. Everywhere I went,
I carried with me a large wicker basket of CASE documents and quilting
materials, and friends started saying, "Here comes the basket lady," when
I came in sight.

That first day, Tom's initial move was a motion to strike the hearing
on the grounds that "this case is not ripe" because PSO did not have a
fuel or water supply contract, could not meet required water quality standards, and had not shown that it could finance the construction and
operation of the project.

Chairman Wolfe denied the motion, but there was a tiny ray of hope.
He said that he hoped water availability would be assured during the
three weeks he was allotting for the environmental phase of the hearing.
He would not consider the hearing satisfactorily concluded until the
water issue was resolved.

Having finished his introductory statements, Chairman Wolfe and
the board listened to limited appearance statements from about eighty
CASE members and supporters. These were short, usually five-minute,
speeches or written statements from ordinary citizens that served the
same purpose as inundating a congressman with cards and letters and
telephone calls. While such statements usually sound totally repetitious
after the first half-hour because there are only so many ways one can say
the same thing, there was the occasional colorful individual, with a striking or especially eloquent way of presenting an idea.

Cathy Coulson-Currin poured cups of drinking water from Oologah
Reservoir for each of the board members, because one of the PSO justi-

fications for using the water was an engineers report that "raw water from Oologah Reservoir will have inherent taste and odor characteristics very difficult to remove by treatment ...[so it] should be used for industrial water needs wherever feasible and ... a new supply of high quality water ... provided for domestic use." Tulsa was already using Oologah water for drinking!

Robert told the board of the report that had convinced him that Black Fox must be stopped at any cost. In the document, Lloyd Mixon, who farmed near Rocky Flats Plutonium Facility outside Denver, told about deformities among his livestock since the plant had been built. Animals had been born without hair or with livers three times the normal size and body cavities filled with bloody fluid. Lambs were born with no bones past the ribcage and a colt without forelegs. Chicks were unable to make their way out of the shells because of deformed beaks and twisted legs.

"Who can truthfully say our unborn children will not have similar deformities?" Robert demanded. "The incidence of deformed children is increasing. Can we say this increase is not brought on partially by the increase in radiation in the environment from bomb tests and releases of radiation from nuclear facilities?"

Gayle Younghein, Ilene's daughter, noted that the Atomic Safety and Licensing Board had only once denied a construction permit, and that solely because of bitter citizen opposition to a plant on the San Andreas Fault. "That doesn't say very much for the objectivity of the licensing process." I was much struck with her observation that any archaeologist, digging up human remains in the future, can tell if they were pre- or post-Atomic Age by the amount of strontium[90] in the bones.

The board merely listened politely the first day. The second day was when they started paying attention, the one in which, from their point of view, they actually got down to business. It was time to present and question expert witnesses.

Vaughn Conrad, a young OSU graduate who worked for PSO, had sat across the table from me at most of the endless prehearings, and as time went on we had become friends, even though we were on opposing sides on the Black Fox question. It was his responsibility to place in evidence the construction permit application,[3] the preliminary safety analysis report,[4] and the six-volume environmental site report.[5] One encouraging moment in the hearing came when Vaughn estimated the cost of site restoration in the event a construction permit were not granted at between one and two million dollars. The very fact that PSO had considered such an issue cheered us—a little!

John Robinson from Black & Veatch, an engineering firm in Kansas, was the first expert witness on the PSAR. One of the questions Tom Dalton asked him was how downstream users of the water would receive prior notification of a radioactive spill. "You could ... get on the telephone ... and tell them you anticipated a radioactive discharge and ... advise them not to use this water until it passed," he responded. The "You could" portion of his answer, of course, made it clear that this was not something he had given much thought to.

Citizens of Broken Arrow, Wagoner and Okay, however, were beginning to give it a lot of thought. They wondered if workers at their water filtration plants would be warned to close their water intakes in time to prevent radioactively contaminated water from entering their water systems. Broken Arrow people were especially worried, because their water intake station was only three miles downstream from the site, and by the time someone got on the phone to their plant, the city's drinking water would already have been contaminated. The Broken Arrow town council asked PSO to move its discharge downstream from their intake station.

Their apprehension was not unfounded. Although the incident was not publicized until November—more than a month after our hearing began—a technician at the Shippingport, Pennsylvania, plant "misaligned two valves, releasing 9,000 gallons of water contaminated by radioactive tritium. It ... poured into the Ohio River." While the spill was dutifully reported to the NRC, plant officials didn't bother to say anything to people in Midland, a mile downstream, and "an immeasurable amount of irradiated water flowed into the town's water system."[6] The NRC was unconcerned, explaining, "that's common until the bugs get worked out."

John West, PSO's manager of Black Fox engineering, made a point of the company's commitment to complying with the NRC's as-low-as-reasonably-achievable regulation. According to PSO, this meant an annual average exposure of 144 millirems, even though nuclear plants operating in 1975 had not been able to bring the average exposure below 500 millirems.

Under Tom's implacable questioning, Dr. Robinson had to admit that the regulations permitted only five millirems per person per year. That an actual average exposure 100 times the permitted maximum did not result in wholesale plant shutdowns was a clear indication that protecting human beings from exposure to radiation was not a serious consideration—either to plant owners or to the NRC.

Another PSO witness was Professor of Radiation Biology Marvin Goldman, Director of the University of California at Davis Radiobiology Laboratory. Dr. Goldman admitted that, "There is no such thing as a 'zero' risk," but said that he believed that the National Academy of Sciences overestimated the actual risk.

Our first witness was Dr. Rosalie Bertell, a senior cancer research scientist at the Roswell Park Institute in Buffalo, who decried "a frightening and growing disregard for life" in the nuclear industry. She cited the unavailability of "statistics on the offspring of workers" eighteen years after the first nuclear power plant had begun operation and the industry's contention, without proof, that "low levels of exposure to radiation are acceptable for man."

Dr. Bertell is a superbly qualified researcher whose statistical and epidemiological work on the biological effects of ionizing radiation was well respected by her colleagues in and out of the government agencies and by members of both houses of Congress. This was not her first testimony before the NRC. Slight and unassuming in appearance, she fits all the stereotypes of mathematicians, but in reality she is a warm, caring human being who is willing to look at the human consequences of technological innovation.

PSO lawyer Rowe subjected her to an extremely hostile cross examination, so much so that at one point Chairman Wolfe felt it necessary to tell Rosalie to take her time before answering. Rowe seemed to be primarily interested in establishing that she was not an engineer, had never worked in a nuclear plant, and that thus—by implication—anything she had to say was beside the point.

He tried to prove that she didn't know what she was talking about because she had difficulty trying to make sense of contradictory NRC standards. Rather than testifying about whether the plant could meet the standards (which was acceptable according to NRC rules), she was challenging the standards themselves (which the NRC was not willing to permit). He ended up asking that almost all of her testimony be stricken from the record on the grounds that she had not supervised all the work that she cited.

Rosalie gamely tried to answer his deliberately confusing questions adequately and managed to make the point that when plant monitor figures for radiation exposure were compared to independent monitors, the plant monitor figures were invariably too low. In the face of Rowe's contemptuous baiting, she struggled to maintain her composure, but at one point it was clear that she had become very flustered.

Joe Howell gave a stirring description of Rowe's character assassination attempt in the *Tulsa Tribune*.[7] "The scene was reminiscent of an old movie thriller in which the heroine was tied to the railroad tracks and a great locomotive was about to run over her. The question appeared to be 'Will she be rescued?' not 'Will her arguments prevail?' "

NRC Physicist Frederick Shon, a member of the licensing board, rode in on his white horse at that point, helping her make the point that no number of millirems of exposure could in itself be used for prediction, and that to determine whether the public's exposure was within the standards, it would be more useful to establish a baseline for the health of the area population. This would make it possible later to test the effects of the radiation that would inevitably be released.

Before he began his cross examination when the hearing reconvened the following day, the NRC's Dow Davis offered evidence that Rosalie's citations of other people's work were perfectly proper. In his questioning, he and Dr. Shon allowed her to explain that the research she was involved in showed that in the long run very low levels of radiation cause genetic damage to the whole population—because, while they produce injured sperm and ova that cause chronic disease, they do not prevent conception—whereas higher levels simply kill people faster, thus preventing them from reproducing the injured genes.

Dow Davis was one of the two NRC lawyers who had tried to persuade me to drop our intention of intervening all those years earlier. A personable, friendly, and altogether delightful young man, he often had lunch with Cathy Coulson-Currin and me during breaks in hearings. His friendship with us did not affect his cross examinations in the slightest: When he went back into the courtroom, he went back into cutthroat mode—so much so that I not infrequently felt that he had taken a stiletto to me.

During our luncheon conversations, we originally stuck to more neutral topics: Where did you go to school? What's that you are making? Do you have any children? Later on, however, when we had gotten to know—and, in an odd way, trust—each other better, Cathy felt comfortable in asking him the question that had been on both of our minds: "As time goes on, how are you going to be able to deal with the fact that through your work you are responsible for so much human suffering and so much damage to the human race?" He had no answer, but when he resigned from the NRC after the Black Fox health and safety hearing, I wondered if Cathy's question had had some effect on his decision.

When PSO lawyer Rowe was allowed to resume cross examination, he had, outwardly at least, modified his earlier attempt to discredit Rosalie as a witness and reduced his demands for pieces of Rosalie's testimony to be stricken from the record. Later in the hearing, however, PSO and the NRC introduced witnesses who flatly contradicted everything she said and whose testimony was accepted without question. I was outraged at the treatment accorded her. Nevertheless, while in the end two of her contentions were denied, the most important one was allowed to stand!

After Dr. Bertell stepped down from the witness stand, the NRC called on Jan Norris to submit the Final Environmental Statement for the record. Tom Dalton was a past master at making what seemed minor points into substantial delays. The FES had been submitted to the Council on Environmental Quality, but they had not yet ruled on it. Tom contended that it had to be approved by them, saying that he couldn't remember which section of the statute said so, but suggesting two possibilities. The NRC people could neither affirm nor deny what he said, so it had to be looked up.

Then Tom wanted to be told the purpose of the FES. Of course, he knew as well as anyone that it was simply one of the requirements of the process, but he succeeded in goading Dow Davis into retorting, "The intervenors have contended that the FES is inadequate. We're introducing it to show that it is not." Tom also succeeded in delaying its admission as evidence on the grounds that it was still under debate.

One of Tom's major triumphs was getting Jan Norris to admit publicly that the fact that there was no place to store the waste from the plant would not prevent the NRC from issuing a construction permit or an LWA.

TW 8-28-77 Editorial

How To Kill A

AT THIS moment, Tulsa's future water, power and sewage disposal needs are all being endangered by a tiny but vocal group of protesters and dissidents.

If this minority of dissatisfied citizens should succeed in killing one or more of these essential programs, Tulsa can kiss its future growth and prosperity goodbye. If you doubt it, just look around at the dozens of U.S. cities that are now dead or dying because they failed to solve these problems.

This need not happen in T' it is unlikely that it the great major' would

nuclear po best hope ply for b people, fore an Regul handf devel rese' sou

Twenty Four

"How to Kill a Turkey"

"How to Kill a City," screamed the headline on the front page of the August 28th *Tulsa Sunday World.* The publisher was so exercised about the Black Fox intervention that he took the almost unprecedented step of printing the editorial on the front page to make sure the greatest possible number of people would see it. "At this moment," he proclaimed, "Tulsa's future water, power and sewage disposal needs are all being endangered by a tiny but vocal group of protesters and dissidents."

John Zelnick and son, Paul. (World Staff Photo)

John Zelnick and son, Paul

If this minority of dissatisfied citizens succeed in killing or permanently stalling one or more of these essential programs, Tulsa can kiss its future growth and prosperity good-bye. If you doubt it, just look around at the dozens of U.S. cities that are now dead or dying because they failed to solve these problems.

This need not happen in Tulsa. Indeed, it is unlikely that it could happen if the great majority of ordinary citizens would make their views known with just half the fervor of the anti-growth, anti-progress, anti-everything minority. But an adequate water supply, a reliable power source and a satisfactory sewage disposal system can no longer be taken for granted.

… Black Fox is under attack by a handful of anti-nuclear, anti-development organizations representing at most a few hundred souls. The "antis" write letters by the dozens to public officials, Congressmen, newspapers and anyone else who might be interested. They religiously attend meetings and make speeches.

That editorial provoked many letters to the editor, pro and con, and at least one canceled subscription.

CASE members Joyce Nipper and Kathy Groshong were so furious, they put a quarter page ad in the following Sunday's *World* to say, "We are tired of being persecuted!!! We care about Tulsa's future and sky-rocketing utility bills." They concluded, "It is not the quantity but the quality of growth that is important to the future health of a city."

They were not alone in their indignation. A young man named John Zelnick was so incensed that he took three days from his job as a technical analyst at Cities Service to write a full page ad, "How to Kill a Turkey." Two days before the *World* editorial came out a tall, handsome stranger entered the hearing room and asked for permission to give a limited appearance statement. He took the stand, said that he was John Zelnick of Broken Arrow, a newcomer to Oklahoma, and told the board that he objected to the fact that PSO would be dumping Black Fox coolant water three miles upstream from Broken Arrow's water supply intake. "If there's radiation leakage, we might all glow in the dark. Gentlemen," he said, "This is ludicrous!"

As John left the hearing room I was waiting for him just outside the door. Extending my hand and introducing myself, I told him what a wonderful surprise his appearance had been. Then I handed him a copy of the Final Environmental Statement, explaining what it was, and asked him to read it. "It looks like a telephone book!" he smiled, as he thanked me and walked away.

It cost John almost $4000 and, eventually, his job, to have his advertisement published in both the *World* and the *Tribune* on Labor Day, the day after Joyce and Kathy's appeared.

Using the word *turkey* in the old Broadway sense of *dud*, John, who holds mechanical engineering degrees from Swarthmore and Penn and an MBA from the Wharton School, wrote a lengthy economic critique of Black Fox in which he concluded that "No private corporation acting in a free market environment would ever adopt the Black Fox nuclear project over a coal capacity alternative" and went on:

PSO probably feels that it is too committed now to consider

killing the Black Fox project on its own. Once large projects such as Black Fox begin rolling, they develop a momentum of their own. The cost analysis work is often obscure and based on someone's best guess, it then becomes difficult to even question the project without offending someone's competency.

Two of his most important points were that without government subsidies the nuclear industry would die tomorrow and that since rates are based upon return on capital, a larger capital base would result in larger salaries for its management. Naturally PSO management would favor building Black Fox!

John announced the formation of his new organization, Citizens Against Black Fox, declaring "If it were not for Mrs. Dickerson, PSO would now be building its 'turkey.' "

When we arrived at the hearing the Tuesday after Labor Day, the whole room was buzzing about John's ad. Doris Gunn and some other CASE members went down the street to the newspaper office for a hundred copies to distribute to our members. They had only five left! But when they learned she was from CASE, they gave her the hundred copies they owed John as an advertiser.

That Friday, John asked that his ad be entered into the record as a limited appearance statement and told the hearing board that his new organization would be polling public opinion and pushing for a referendum on Black Fox.

Over the years, John became more and more adept at using his charm and speaking ability on behalf of our cause, and two years after the hearing he decided to enter the 1980 senate race.[1] Characteristically, instead of using a news conference John paid $1000 for a three-column ad in the *Tulsa World* and *Tribune* to announce his candidacy. Thus he could be sure that any distortion of what he stood for would have come from his own pen rather than those of reporters. John knew that his opposition to nuclear power would put so much money into the pockets of his opponents that he had little possibility of actually winning, but he felt that the opportunity for added publicity and public debate on the issues made the effort worthwhile.

One of the more colorful characters CASE brought to the hearing was Dr. Charles Hyder of Albuquerque, a former NASA employee with a doctorate in astrogeophysics. Tall and broad, with his hair falling from a bald spot to his shoulders, Charles did not have a meek bone in his body. Like Tom Dalton, he knew his own mind and was not afraid to speak it. They got along like a house on fire, but Charles was not deferent enough for the PSO and NRC people. In fact, he was not deferent at all. It was readily apparent from their expressions that they would like to light a fire under him!

Charles Hyder's father had made a fortune in uranium mining. By the time that fortune came into Charlie's hands, however, the costs and suffering caused by nuclear bombs and reactors had become so obvious to him that he had come to think of it as blood money, and he gave it all away!

Not a person to suffer fools gladly, Hyder taunted PSO with the fact that private insurance companies were not willing to cover the nuclear disasters that nuclear proponents said could not happen. He declared that, "Until you know what to do with the wastes, I don't believe you are entitled to produce them." He ridiculed former AEC Chairman James R. Schlesinger's bizarre suggestion that we shoot high-level radioactive wastes "into the sun—taking them right out of the world."

His testimony was bulky and covered subjects ranging from uranium supply to the philosophy of the NRC's hearing process. Bob Mycue, of the *Tulsa World* commented during a recess that he must be an expert in all fields. "Yes," he agreed, "Some men know about the trunk, the tusks or the hide of an elephant. I know about the whole elephant."

Hyder's opposition to the nuclear industry cost him his job with NASA as well as a lectureship at the University of New Mexico. After suffering under the lash of Hyder's tongue, PSO's attorneys and the NRC staff were itching to give him his comeuppance. There was considerable wrangling over his testimony, and they managed to get large portions thrown out.

The fourteen days of the hearing in August and September often seemed endless. Tom Dalton studied the witnesses' testimonies from beginning to end and stretched out the hearing by questioning minute—and often hum-drum and monotonous—details. On occasion, the witnesses themselves produced electricity in the courtroom, bringing an atmosphere of color and drama to the proceedings, but more often we would sit there wondering how to keep awake until we had another opportunity to enjoy Tom's droll wit. Tom was never boring, but the others frequently, and I suspect deliberately, were!

The dark side of the anticipated boost to Inola's economy was exposed by OU geography professor Edward Malecki. He pointed out that it would be a boom-and-bust situation in which a temporary doubling of population would cause enormous problems. Expenses for schools, transportation, and public services would not be covered by new taxes until 1985, considerably after they were no longer required.

Professor Robert Halvorsen flew down from Seattle to discuss projected power demand. That evening when the hearing regulars got together at the Middle Path Cafe, he told us quietly, "The economics will kill nuclear power."

As one of our expert witnesses, OU Professor Karl Bergey urged PSO to lease or sell wind-power collectors to home and business owners and bring their excess electricity into the power network. Later, he began a new international career, with his son Mike and daughter-in-law Jan, as a windmill consultant and contractor. Mike and Jan would come to Tulsa to put on a wind power seminar for us in 1992.

All of these issues were interesting in themselves, but they were things that, for the most part, we had already been discussing for months or years, so they held no excitement for us. It was often more stimulating to see them through the eyes of Joe Howell in the afternoon newspaper, the *Tulsa Tribune*, and of Bob Mycue in the morning newspaper, the *Tulsa World*.

One year shortly after World War II, I taught home economics at the Chelsea high school, twenty miles from Claremore. We had only one car, and I had not yet learned to drive, so Robert took me to the bus station every morning to catch the Greyhound for the trip up Route 66 and then returned to bring me home every evening. (I had plenty of exercise that year, because the high school was a mile from the bus station and I walked that mile morning and evening!)

The bus was often late, and as I sat in the waiting room, there was usually a man with a businesslike expression and a twinkle in his eye waiting for a bus going in the opposite direction, to Tulsa. His name was Joe Howell. With his sweet-faced wife, five daughters and young son, he lived at the corner of Third Street and Seminole in Claremore in a big white house surrounded by a tall box hedge. His wife, Mary, spent much of her time as a Girl Scout leader. Every day he rode the bus to Tulsa to write his stories for the *Tribune*. Joe's daughters went to high school with mine when I taught home economics in Claremore in the 1950s.

As I continued my fight to defeat Black Fox, Joe's stories chronicled our triumphs and defeats, our convictions and our uncertainties. His stories were always scrupulously evenhanded. He never editorialized; he always reported. Like Tom Dalton, he had a knack for going to the heart of the matter and bringing it alive for the reader. Nothing that I read in his articles ever gave a hint that he was actually a Black Fox supporter.

Our conversations during lulls and recesses in the hearings, however, were a running and frequently impassioned debate. Throughout the hearings, he was there every day, often with his wife, and you could always tell when the hour of 11 a.m. drew near because that was Joe's deadline.

I had met Bob Mycue early in the hearing process and discovered that he was very definitely anti-nuclear. He would send me copies of anything that came across the wires about energy that he thought would be helpful to our cause, and we often discussed things that happened during the hearings.

Bob was especially good at drawing out expert witnesses and often added human interest stories that could not be brought out in the hearings. One day he came into the courtroom fuming and told me that the evening before, his editor had pitched his story into the waste basket. "If you can't find something good about PSO to write about," he had been told, "don't write anything good about CASE." And Bob apologized about the tone of that morning's story.

Bob was ecstatic when Black Fox was canceled, and when, three or four years after the cancellation, he died of lung cancer, I was comforted to know that he had lived to see that day. His wife called to tell me that he had arranged for his obituary in the newspaper to ask that his friends send contributions to me in lieu of flowers. I was grateful for those contributions and deeply honored that Bob wished me to receive them.

A third newspaperman gained my especial respect and gratitude during all our vicissitudes and triumphs. Frosty Troy received his seasoning as a reporter in his thirteen years on the *Tulsa Tribune*. Then in the prime of his career, he left to establish a weekly news magazine in Oklahoma City. Some of the most important articles on the Black Fox fight were published in the *Oklahoma Observer*. Frosty and his wife Helen, as its publisher and editor respectively, had the independence to encourage our work against Black Fox, but the result of their principled stand was that PSO and, later, Oklahoma Gas and Electric cancelled their ads in the *Observer*. Amusing and articulate in person, Frosty provided a much-appreciated point of view when he had occasion to speak to raise funds for CASE's legal costs. The *Observer* continues to offer a respected forum for civilized and controversial discourse on any issue that affects the citizens of Oklahoma.

Tom Dalton subjected one of PSO's local "experts" to a catechizing that mirrored PSO's treatment of Rosalie Bertell: The hapless Steven Day was not familiar with a study done for the City of Tulsa which concluded that by 1983 the City of Tulsa would be out of water, with another study performed by the Corps of Engineers, or with the "Oklahoma Water Plan." Nor did he know anything about the Corps of Engineers study of the Verdigris River basin water supply and water quality, or of the water supply problems of Broken Arrow and Sapulpa. He was even unaware of the Oklahoma Office of Community Affairs and Planning water study!

There was considerable testimony and discussion on coolant water for the plant. NRC witnesses made it clear that the Verdigris River, by itself, could not support the plant and that Oologah water would have to be available. Another testified that mercury concentrations in the Verdigris were already so high that there would be times when the plant would make them exceed EPA standards. Two young India-born water experts, Umesh Mather and Jim Shirazi, testified—Umesh for CASE and Jim for the Oklahoma Water Resources Board. Both agreed that the minimum Verdigris River flow given in the FES was too high by a factor of between three and five.

Umesh, like Jimmie Pigg later on, severely criticized the portion of the environmental impact statement that discussed heavy-metal pollution and living creatures in the Verdigris River. The board members seemed not to be very concerned about any of this until Jim Shirazi and James Long of the Oklahoma Water Resources Board asserted that PSO could not be given a permit without a public hearing.

Shirazi warned that OWRB standards were oftentimes more stringent than EPA's, and that if and when a permit was issued, PSO would not only have to comply with them but that OWRB would review their compliance!

Negotiations on the water contract stalled, and the board took a five-week recess on September 10, 1977, the day they had planned to complete the hearing.

Carrie at the Tulsa State Fair, 1974

Fairs and Energy Expos

Every year during our fight to stop Black Fox, CASE had exhibitor's booths at the Tulsa State Fair and the Rogers County Fair, and the Sierra Club joined us in our booth at the Tulsa State Fair during the latter years of the hearings. The fair booths provided a means of getting our message to the public about the problems of nuclear power and a place to sell the items we used to raise money.

A young CASE supporter named Ron Surface operated a Tulsa store that sold survival foods. For several years, he organized an energy exposition on the Tulsa State Fairgrounds each year. The first Energy Expo was held that September, 1977, and lasted for three days. Ron was a superb organizer. He would start months in advance discussing ideas for exhibits with potential participants, arranging space and support facilities, and plotting out the logistics of each event. The Energy Expos were always preceded by a tabloid insert in the *Tulsa World* showcasing exhibitors, and even the tickets were printed a good three months ahead.

Ron always managed to present a fascinating combination of exhibits to

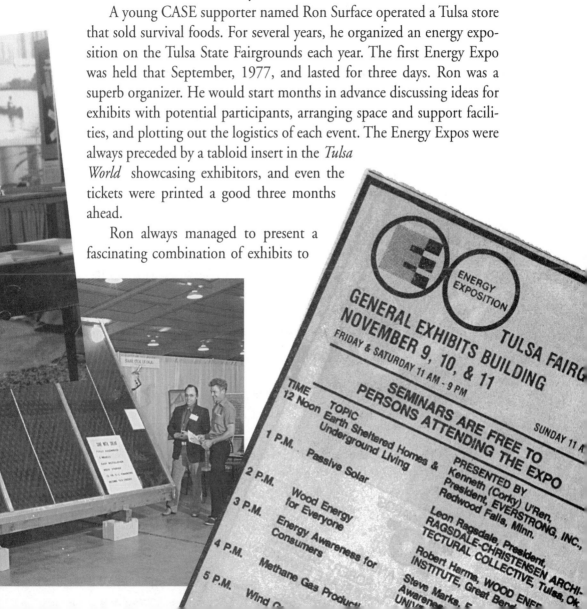

ENERGY EXPOSITION

GENERAL EXHIBITS BUILDING
NOVEMBER 9, 10, & 11
FRIDAY & SATURDAY 11 AM - 9 PM
TULSA FAIRG
SUNDAY 11 A

SEMINARS ARE FREE TO
PERSONS ATTENDING THE EXPO

TIME	TOPIC	PRESENTED BY
12 Noon	Earth Sheltered Homes & Underground Living	Kenneth (Corky) U'Ren, President, EVERSTRONG, INC., Redwood Falls, Minn.
1 P.M.	Passive Solar	Leon Ragsdale, President, RAGSDALE-CHRISTENSEN ARCHI-TECTURAL COLLECTIVE, Tulsa, OK.
2 P.M.	Wood Energy for Everyone	Robert Herms, WOOD ENER INSTITUTE, Great Ben
3 P.M.	Energy Awareness for Consumers	Steve Marks, E Awareness UNIVE
4 P.M.	Methane Gas Produc	
5 P.M.	Wind G	

pique the interest of apartment dwellers, homeowners, and potential home-owners. A windmill was set up on the fairgrounds to provide power for a local radio station to broadcast from the Expo. One year, a Tulsa contractor constructed an envelope house. It was so well insulated and collected heat so efficiently in the winter—much like a greenhouse—that it was practically unnecessary to provide any artificial heat source at all. At nearly every Expo there was a man who traveled around in a truck onto the back of which he had built a little wood-shingled house fitted up with every energy-efficient gadget that struck his fancy; I think the truck may even have run off methane.

Ron himself always had a booth filled with back-to-the-land books, and there were solar collector and window-film exhibitors, wood-stove exhibitors, chain-saw and wood-splitting exhibitors, natural-fiber clothing exhibitors, a solar water-distiller exhibitor, food-dehydrator exhibitors, and exhibitors for all those back-to-the-land technologies Ron's books talked about. Other exhibits highlighted insulation, recycling, and conservation. There was even an exhibit of orgone boxes (although they were called pyramid power collectors). And there were always exhibits from Oklahoma Natural Gas Company, PSO and, of course, CASE!

I always looked forward to fairs and expositions, because they offered me an opportunity to spend some time with my granddaughter Sandra and my "adopted" granddaughter Camille, preparing posters for the booths. When I was in nursing school, I had developed a friendship with Camille's parents, Eunice Myall, then the librarian at the Hillcrest Medical Center School of Nursing in Tulsa, and her husband Dick. Camille had no living grandmothers, so, I "adopted" her. Those years of the struggle against Black Fox were also the years when my grandchildren were growing up, and one of the things the campaign stole from me was my time with them, so those times when I could be with my grandchildren were doubly precious. I reveled in the moments I could spend with the two girls and Jim's children, Melissa and J.J., when they helped me fold letters and stuff and stamp envelopes.

People came to the Energy Expos from miles around. There was a warm, happy holiday atmosphere, as young people thinking of setting up energy efficient households mingled with families and grandparents.

Every year some of the booths at the Expo were reserved for area schools to show off their projects—somewhat like a science fair. That year, one of them was occupied by an exhibit of the work of Paula Baines' class. Her students had completed several solar and energy conservation projects, and she and some of her students were always there answering questions about their work.

Several years earlier an elderly Chelsea woman named Mattie Bains had moved into Aunt Carrie's Nursing Home. Her son, Frank, and his wife visited her quite often. Frank's wife was a Chelsea schoolteacher named Paula, and Frank had been one of my home economics students when I taught at Chelsea.

In the late forties boys just didn't take home economics, and for a girl to be in a shop class was nothing less than a scandal. How did Frank get to be one of my home economics students? Well might you ask!

The first day of the school year when I taught at Chelsea, three boys came to me and asked to enroll in my home economics class. Thinking that they were just trying to cause mischief, I agreed that they could—if they received permission from the superintendent of schools. Presently they came back to tell me that they had the necessary permission.

For the next month, they baked, sewed and studied the same curriculum as the girls in the class. Their biscuits and cookies were mouth-watering, and they were inordinately proud of the meals they prepared. They had all learned to measure fabric and use a thimble, and the embroidery on their tea towels was perfectly acceptable. I would go home and proudly tell my family what a progressive superintendent of schools we had in Chelsea, a man who understood that it was just as important for men to know about household tasks as for women.

A month after they joined the class, the superintendent walked in unexpectedly and looked around to see what we were working on. I was preparing to show him proudly that the boys had been able to accomplish just as much as the girls, when I heard his voice demanding dumbfoundedly, "Whaaat are these boys doing here?" That was the end of their adventure and the end of my "progressive" superintendent.

In the CASE booth that fall, the *Sun in Splendor* quilt made an impressive backdrop. It seemed to glow, even without help from the lights. One man fell so much in love with it, that he took a hundred dollars' worth of tickets. On the counters were displays of silk-screened T-shirts and aprons with conservation or alternative-energy motifs, Cathy Coulson-Currin's Red Fox prints, and a variety of buttons, pendants and bumper stickers. Paula stopped by and bought over a hundred dollars' worth of T-shirts and other odds and ends for her grandchildren.

She had packed them into the bag she brought with her and was about to leave when I asked her if she wanted a chance on the quilt. She handed me a dollar bill. "I'll take one chance," she replied, "and if I'm supposed to win, that will do it." She filled out her ticket and dropped it in the enormous bin we had provided to hold the 3000 tickets.

The third day of the Expo was drawing to a close, and the radio station set up beside our booth was preparing to broadcast the names of the winners of the various drawings. We turned and turned our tickets to mix them one final time.

Robert picked up Cathy Coulson-Currin's little two year old daughter, Tracy, and asked her to close her eyes tight and reach over and pick up one ticket. The ticket read, "Paula Baines."

Full of mischief, I hurried over to Paula's booth and asked her to announce the winner of the sunburst quilt over the radio for me. "Find someone else, Carrie," she said, "I'm starved."

The announcer was waiting. "Come on Paula," I coaxed. "It will only take a minute, and I'll hold your lunch while you announce it." Handing her sandwich to me, Paula read the name of the winning ticket into the microphone. "Paula Baines."

"Paula Baines! That's *my* name," she gabbled, stunned. "I can't believe it. I must be doing something right. I won the sunburst quilt!"

Tom Dalton's Bombshell

The world did not stand still while we waited for the hearing to resume. The Chinese government detonated a four megaton bomb on September 17, and six days later the cloud of radioactivity had crossed the Pacific and was halfway across the United States. This is one of the few times we can be grateful we were having a minor drought. If there had been rain, it would have brought the fallout to earth on top of us.

A few days later 30,000 protesters defied a force of nearly 20,000 police to march on a heavily fortified experimental fast breeder reactor at Kalkar in West Germany.

Nuclear Foes Say They Risk $1 Million

Opponents of the proposed Black Fox nuclear plant have a response for Public Service Co. of Oklahoma officials who have already ... $30 million in thei- ... Oppo-

group which has bee- zens Actio-

SECTION

AY, OCTOBER 22, 1977

Water Supply Question End Black Fox Hearing

Questions hanging over the water supply arrangement for Public Service Co.'s planned Black Fox nuclear plant continued to be argued Friday as the hearings closed.
A spokesman for the Nuclear Reg-

been in negotiations for four years. Asked why it has taken so long, Dwen blamed "administrative delays with-in the corps."
Dwen's testimony also covered the ratifications of a League of Women

BLACK FOX environm site suitability hearings la a month, beginning Aug. 2 three weeks, and reconver final sessions this week.
If, after considering th

Five days before the hearing resumed, the *Wall Street Journal*, usually a bastion of *laissez faire* economics, carried a long article describing the enormous danger, difficulty and expense of dismantling a nuclear reactor. All I could think of as I read it was, *What a dreadful waste of time, energy, money and precious natural resources! There is no way that renewable energy could ever cost more than nuclear power!* I suspect that I was not the only one who reached that conclusion.

That same article noted that the White House Council on Environmental Quality had just recommended that the NRC stop issuing licenses for new nuclear power plants until "acceptable ways are found to dispose of radioactive wastes generated by the plants while they are in use, as well as acceptable ways to clean up the plants after they are abandoned."

"Deadlines should be established for solving the waste disposal problem," its chairman urged. "Licensing of new nuclear plants should be halted if no solution has been found by the proposed deadline." If the NRC had followed those recommendations, Black Fox would already have been out of the picture!

Ten days before the hearing reconvened, the controversial water contract between the City of Tulsa and PSO was finally signed. We feared that we would not be able to hold off the LWA for more than a few days.

When the hearing resumed on October 17, we started pushing the issue of escalating capital costs of nuclear reactors and the comparative attractiveness of practical alternatives. Mike Males, who had moved to Montana to work for the National Park Service, came back to act as an expert witness on these issues.

Another of our witnesses, Jimmie Pigg of Moore, Oklahoma, gave a discussion and critique of the biological portions of PSO's environmental report. A biologist for the State Health Department and science coordinator for the largest school system in Oklahoma, he had also worked as a marine biologist for the Navy and trained Peace Corps fishery biologists. Referring to the canaries that were once taken into coal mines because their deaths warned miners that they were in danger from lethal gases, Jimmie called the fish in the Verdigris "canaries with scales."

Jimmie Pigg charged a very nominal fee, but because of all the other fees, I had to postpone his payment so long that Ilene paid it out of her own pocket. When I ran into Jimmie at a state environmental meeting in 1992, he hugged me. "Carrie," he declared, "If I had known you and Robert were sacrificing so much, I would never have charged you a cent!" The warm little glow that left in my heart was a balm to my bruised spirit.

As I walked into the meeting room on the last day of the hearing, Tom Dalton pulled me aside with a smile that seemed to be embellished with canary feathers. "I have a big surprise to spring today," he murmured. Knowing how much he loved drama, I didn't ask him for details, but all through the day I was on needles and pins, wondering what it would be.

One of PSO's Assistant Vice Presidents had attended the conference at Sudye's house. In his testimony at the hearing he dismissed solar and wind generation as serious contenders in Oklahoma because of their "cost, technical immaturity and unknown environmental impact."

How could he be so blind? I asked myself. I wanted to get up on the witness stand and shout, "Let's use our good old American ingenuity! If we could complete the Manhattan Project, we could do even better on a Sun and Wind Project!" That impulse was futile: Not only was I not an expert witness, as co-chair of CASE I was not even allowed to make a limited appearance statement!

I suspect that even then there were sensible people in PSO saying to each other that they had to think of alternatives to nuclear energy. PSO engineers had ready for the Department of Energy a proposal for a 200-acre heliostat field near Oologah to harness the sun's energy,[1] and their 1979 annual report indicated that this was only one of several alternative energy sources they were investigating. We never learned whether the DOE had approved or disapproved PSO's solar enterprise. It was never built. PSO was also one of fifteen electric utilities under consideration for demonstration wind-turbine systems to be tested by ERDA.

An interesting illustration of the ways in which legislation and regulation can deceive us all can be seen in the Power Plant and Industrial Fuel Use Act of 1978. For some reason, possibly misinformation from oil companies seeking to manipulate the market, Congress was convinced that we were about to run out of natural gas, a much cheaper, more readily available energy source than oil. To conserve what little gas they thought we had left, they included in the act an order to limit the use of natural gas severely in power generation.

One summer during the late 70s, Patricia was driving cross-country from Massachusetts to Oklahoma, when for the first time she happened to be crossing southern Indiana and Illinois after dusk. The next day she expostulated, "Mother, I thought you told me we were running out of natural gas!"

"Well, that's what Congress says," I assured her.

"In that case, why are they burning it off the oil wells in Indiana and Illinois?" she wanted to know.

I had no answer for her, so she went to her brother Jim and asked the same question. "Oh, it's just that when they drill for oil, there's always natural gas on top, and it's so plentiful and cheap that it's not even worthwhile for them to bottle it, so they burn it off!"

Of all the fossil fuels, natural gas produces the least pollution, and the realization that it was being squandered for no reason was hard to live with. Eventually, Congress was undeceived, and the Power Plant and Industrial Fuel Use Act of 1978 was repealed. I am glad to know that natural gas is once again recognized as a viable energy source, but I shudder to think of the other ways in which unbridled profit motives warp our lives and waste irreplaceable resources.

As the last scheduled witness for the day gave his testimony, people were gathering up their belongings preparatory to the long-awaited adjournment of the hearing, but I was sitting in suspense, wondering if Tom Dalton had forgotten the surprise he had promised me.

Then out of the blue, Tom said in a conversational tone of voice, "The city does not have any water, and without the Oologah water there is no water available for Black Fox Station."

The hearing officers looked dumbfounded. After all, the City and PSO had signed a water contract!

The room was silent. People forgot to breathe. All eyes were focused on Tom Dalton. "The city does not have a contract with the Army Corps of Engineers for storage of water in Oologah Reservoir. A 1958 contract between the city and the corps was terminated in 1961 and was to be renegotiated, but the renegotiated contract has not been completed."

Tom had called the Corps of Engineers office months earlier and discovered all this, but he had not breathed a word of it until that minute!

PSO's Miller was desperate. "It is highly irregular," he shrieked, "and I object." Instead of the expected adjournment, the hearing was recessed,[2] and Jim Dwen of the Tulsa corps district's legal division was asked to help establish the facts of the matter. Dwen confirmed what Tom had said but reiterated that they were trying to complete the contract.

Re-negotiations had begun when Oologah Lake was enlarged in 1973, and Tulsa's storage space in the lake was to be expanded by a factor of ten. Without the larger allocation, Tulsa could not fulfill its agreement to provide water for the new coal-fired plant and the Black Fox plants. Dwen admitted that negotiations were incomplete because of administrative delays within the corps. He said that the League of Women Voters' pending request for an injunction to require an environmental study on the contract was also a factor.

With that on the record, the hearing was adjourned, and attorneys from the three parties were required to present their "findings of fact" to the hearing board. If after they considered the evidence their findings were favorable, the board would authorize the NRC's staff to issue PSO's long-awaited limited work authorization.

I was so proud of Tom Dalton. He had taken the initiative to search out evidence available to anyone, but unknown even to PSO—who would have been expected to leave no stone unturned to achieve their goal. Because of his intelligence and industry and ingenuity, the hearing had lasted four weeks instead of the three planned by the board and had ended five weeks after its planned adjournment date.

Those extra weeks had not only gained us maneuvering room, they had given us more publicity than we could ever conceivably have bought. Public awareness and debate on the issues we were raising had increased to a degree we could never have predicted when we began.

It was terribly difficult to raise enough money to keep Tom going, and, as in the past, it was often necessary for Robert and me to write personal checks to make up the difference, but Tom Dalton had done us proud! He was worthy of his hire.

Years earlier Cathy Coulson-Currin and I were driving down Peoria Street in Tulsa, when we saw a billboard in front of a hardware store. The billboard read, "What the mind can conceive and believe, it can achieve!" I said, "Cathy, I have to write that down." I pulled into the parking lot, and wrote the quotation on a card which I placed in my purse so that each time I opened the purse it would be visible. I wanted to engrave it onto my brain.

I could not fool myself into believing that the NRC's determination to license Black Fox had diminished, but I would not allow myself to be discouraged by the thought of that LWA. "What the mind can conceive and believe, it can achieve," I reminded myself. I had conceived the goal of stopping Black Fox. I would continue to believe that it was possible. And we would achieve it!

Vaughn Conrad was left with the task of gathering up all the books and materials used during the hearing and packaging them up for shipment back to Washington. As he set to work, I was waiting for Mary to come and pick me up, and we started one of those idle conversations that come in the aftermath of great emotional and mental exertion.

"Mrs. Dickerson," he asked quietly, "do you ever envision a time when you will accept nuclear power?"

My reply was instantaneous and unequivocal. "Never!" I replied. "Even if nuclear plants could somehow be made safe, they would continue to create radioactive waste, and there can never be a satisfactory way to dispose of that. It is morally wrong for us to maim all future generations so that we can enjoy electricity in the present."

Now, with the completion of the hearing there was time to attend to other matters. Now that I no longer had to drive to Tulsa early each morning, I could spend more time with Robert and our children and grandchildren. And I could write fund-raising letters.

PSO made another bid for public sympathy, announcing that they would lose the $30 million that had already been spent if the final permit were withheld. "That's a terrible thing to contemplate. If it happens, we'd just be out everything we've spent up to now," one of their officials moaned, going on to complain that, "The NRC has nuclear plant builders over a barrel, and can pose any number of restrictions before a final permit is issued."

John Zelnick gave a new dimension to PSO's complaining when, only partly tongue in cheek, he told reporters that we had "committed the equivalent of $1 million in money, time and energy in opposing the Black Fox nuclear plants. We could lose every cent of our investment if a final permit is issued for the project."

PSO's announcement was, of course, preliminary to another request for an increase in electric rates. PSO was established in 1915, and in the first nearly sixty years of its existence there had never, ever been a request for a rate increase. Then the idea of Black Fox came along, and everything changed. They suddenly needed a lot of money.

In 1975, the Oklahoma Corporation Commission granted PSO a $15.6 million increase, and only two years later they were asked for another $17.4 million. Tom Dalton represented the Sierra Club and CASE in that rate hearing, and the OCC decided PSO needed only $6.4 million. Commissioner Jan Eric Cartwright spoke for the majority of the OCC when he chided PSO for overbuilding its Oklahoma facilities at the expense of its rate payers. Miffed, they decided to ask for another hike later in the year.

Those of us who were beggaring ourselves to oppose their irresponsible behavior were understandably indignant to hear people who had not risked a penny of their own money demanding the right to rob the rate payers to support their new plaything. Why, we wondered, are not more rate payers up in arms about what was being done to them?

Overruling the Board

For PSO to draw the required amount of water from the Verdigris River, they needed to have in hand a contract for water from the City of Tulsa, a water allocations permit from the OWRB, and an NPDES discharge permit from the EPA.

The water allocations permit from the Oklahoma Water Resources Board would authorize PSO to draw 44 million gallons per day of water from the Verdigris River to cool the reactors. The hearing on that permit was held on February 23 and 24, 1978. The weather was cold and the roads slick.

William Lorah of Wright Water Engineers in Denver testified for us, bringing out two important and related points that would have ended the matter then and there, had the OWRB not been insistent on giving PSO what they wanted and had the NRC not already decided that Black Fox would be built.

The first was that there is a water compact between Oklahoma and Kansas that gives Kansas control of all the water running into the Verdigris River in Kansas. That meant that during a drought the riverbed would be dry when it crossed the Kansas-Oklahoma border. This is no *pro forma* agreement; there were several times in the fifties when that is exactly what happened. If the demand for water in Kansas were to increase because of population pressures or, for instance, another nuclear power plant upstream, that condition could become permanent.

Eighty-one percent of the water flowing into the Verdigris River comes from watersheds in Kansas. If Kansas needed to exercise its rights, only the 19% of the Verdigris' water that came from northeastern Oklahoma would be available to Oklahomans. And in times of drought, that would be a tiny amount.

The second point was that of the 28,000 gallons per minute PSO proposed to take from the river, 90% would be lost through evaporation. In other words, while they talked soothingly of simply using the water and then returning it to the river for the use of people downstream, in

reality more than 25,000 gallons per minute would be lost to people downstream. Many towns and water districts were, at the time, using water from Lake Oologah and the Verdigris River, while others in need of more water had no assurance of a source.

Umesh Mathur pointed out that PSO was requesting twice as much water from the Verdigris River as they needed for cooling so they could dilute the tons of poisonous chemicals they would be pouring into the Verdigris enough to comply with the Oklahoma water pollution standards.

When OWRB members, who were our public servants and were supposed to be looking out for our welfare, fawned on the PSO and NRC staff and treated us as trifling annoyances, it was hard for CASE members to contain their anger and frustration. I sent out a letter asking that each member write a letter to the board telling them how they felt about the water supply situation and asking that the permit be denied. "Keep your comments on a high plane," I urged. "Don't let antagonism or recrimination creep in. It might make you feel better momentarily to get it off your chest, but it will hurt our cause to show animosity."

Our efforts went for naught, and the Oklahoma Water Resources Board issued the permit on July 8, 1978. Tom immediately appealed the ruling to a higher court.

Umesh Mathur

CASE files suit here seeking to block PSO from Verdigr

Citizens' Action for Safe Energy, Inc. — opposing the proposed location of Black Fox nuclear generating plant in Inola — is now suing Public Service Co. of Oklahoma in district court to block its planned

...recognized a beneficial use of water...not termed waste." The board called CASE's "claim that the applicant (Public Service) failed to demonstrate safety requirements for the total (nuclear) system is irrelevant...at this time." It also called CASE's claims of water

to Public Service. It claims that the board have the power to allocate supply of which is govern state Compact." CASE is also currently Aug. 8 water board order city of Tulsa's right to feet of water from Ool It opposes the order in

In the process of producing electricity, the water used in Black Fox would have accumulated high concentrations of pollutants, especially sulfates and heavy metals, and the cheapest way of dealing with this was to dump them into the river along with the hot water. Before PSO could proceed with that kind of disposal, however, they needed a National Pollution Discharge Elimination System permit from the federal Environmental Protection Agency. To get the permit, they had to go through another hearing process which, just as with the earlier ones, would normally have been *pro forma*.

Shortly after the OWRB hearing, CASE's board members met in Tom's office to decide whether to participate in the EPA hearing. Umesh Mathur, our expert witness on water pollution, and Tom Dalton discussed with us the reasons for participating and the costs to be anticipated. We were already in debt to the hilt. Board members reasoned that the cost would be prohibitive, we could not possibly raise enough money, and since there was no possible way we could win, why try?

Blindly shuffling the papers in my briefcase in the shamefaced silence that followed, I tried to conceal my tears of disappointment and frustration, while the board members dispersed. By the time they were all out the door, I had regained control of my emotions, and I knew what I was going to do.

Tom and Umesh, sitting with downcast faces, were trying to avoid each others' eyes, and I cleared my throat to attract their attention. "Tom," I asked, "would it be legal for me to overrule the board if I were totally responsible for raising the funds?"

Both of them rising uncontrollably in their excitement, their faces split in identical grins as Tom exclaimed, "Yes! It would be!"

The change from despair to jubilation left them nearly beside themselves, and I prepared to leave lest I found myself dancing around the room with them! In my presence, no other board member ever referred afterward to that meeting.

When the two-day EPA hearing began on March 22, there were again a number of people waiting to give limited appearance statements. Most of them said pretty much the same thing, although Marilyn

Oldefest sang her testimony to the accompaniment of her guitar. Ilene, Joyce Nipper, and many others had their say, but Nancy Perreault put it most succinctly and forcefully, "Our water is already polluted, and in short supply."

John Zelnick appealed to the EPA for objectivity in the face of all the pressures arrayed against it. "Our lives depend on it," he pleaded.

The NRC Safety and Licensing board had already concluded that, based on the design of the facility, Black Fox would violate state standards when the Verdigris was low. Our witness, Umesh Mathur, felt that the draft permit would give PSO *carte blanche* to evade federal and state regulations that required them to install equipment to clean up the waste water before it was put into the river.

He charged the EPA with disregarding its own regulations and violating federal law in drawing up the draft permit and urged its officials to require more effective measures.

EPA's attorney, Mike Gibson, denied everything, but PSO's Vaughn Conrad was actually more in tune with reality. He conceded there were problems but insisted that "technology is there to comply with EPA standards." Then he put in words what we had known all along was all that the EPA and NRC were paying attention to: "We're approaching the point where we may have to build a huge water treatment facility at Black Fox. This has the potential of running up plant costs."

An NPDES permit—which would allow PSO to get away with constructing the plant without pollution control facilities—was granted on July 19, 1978. CASE immediately petitioned the EPA for an adjudicatory hearing in the matter.

Twenty-Eight

"I didn't get a note!"

Early on in the struggle to raise the funds to defeat Black Fox, I established a routine of duplicating and mailing a letter to CASE contributors every few weeks. I would make a point of scribbling a little personal note in the margin of each letter—at least a "Thank you," or "I hope you can come to the next CASE meeting," if I didn't have time for more. One of our young members was a faithful contributor, and when he did not respond to a letter I had sent, I started to worry. When, after several days, I had not heard from him, I called to see if he were all right, only to find that he had not replied because there was no personal note on his newsletter!

After that, although it took many hours to write them, every contribution usually called forth a personal, handwritten "thank you" along with a receipt. Not only did the notes express my true sentiments, I had found that they were worth the time: The sooner I responded, the sooner I would receive another contribution.

One day I discovered that this tactic also had its drawbacks: I received a small check from a lady in Oklahoma City, with a note pleading, "Please do not send me any more receipts or thank-you notes. I feel guilty if I can't send another contribution right away, and I just can't afford to send them that often."

BLACK FOX

REACTOR SALES:
UP AND DOWN

Nuclear Power
Plants Purchased
by U.S. Firms

35

26

20

14 16

7

4 3 0

1969 1970 1971 1972 1973 1974 1975 1976 1977

LOOKING AHEAD
units must

Daily Oklahoman
Oct 6, 1977

Radioactive Powde

SPRINGFIELD, Colo. (AP) — The spilling of 42,000 pounds of radioactive powder en route to Oklahoma last

processing as fuel for nuclear generating stations.

"They've been transporting material in those drums since day

"They Some are fl ble."

Paul gional

After the EPA hearing, I had reached the point of desperation about all those notes, and in my June first letter to contributors, I threw myself on their mercy, explaining that "the intensity of the situation during the past few months has prevented my having time to express my gratitude to each of you personally." Telling them how much their notes and contributions meant to me, I continued:

> It is sad that a cause is necessary before one can become acquainted with such wonderful people, but it's good to know that from all across Oklahoma, citizens have rallied to the need! Not only are local people from Inola, Claremore, Tulsa, and other nearby localities concerned, but people are responding from such far-away Oklahoma cities as Boise City in the Panhandle, Altus in southwestern Oklahoma, McAlester, Tahlequah, Bartlesville ... Without your financial support the Limited Work Authorization would already have been issued.
> We know the NRC is planning to grant an LWA this summer sometime—but we do not plan to reduce our vigil one iota if and when they do. If it had not been for government subsidies, nuclear power could not have gotten off the ground in the first place. If we can hold out long enough we can win this fight!

I ended my letter by urging them to attend the Radon Gas hearing that would begin the following Monday.

Months before, when I came to understand the importance of delay in stopping Black Fox, I begged Tom Dalton to leave no stone unturned in

ll Called Bigges

d.   A seven-member material from sprea
s cleanup crew worked ing, said Richard
a- inside a specially con- Hornsby, Exxon's envi-
structed shelter de- ronmental coordinator
re- signed to prevent the at the site.
ffi-
vi.

searching for avenues of delay. It should go without saying that putting off the final reckoning was only one of the impetuses for this tactic: Every delay provided more opportunity for people to come to their senses; every delay gave us more time to get the message to people; and every hearing gave us a forum for doing so.

As I knew he would, Tom followed through. Superbly!

The radon hearing was to become an avenue for publicizing one of the greatest tragedies perpetrated in the whole blind rush to force nuclear power on the American people. The fact that the LWA was delayed was simply a bonus. And without Tom's ingenuity and perseverance that would never have happened.

Many years ago, my daughter Patricia and her husband bought an old farmhouse with an attached barn. A farm girl herself, Patty found the idea of letting her children grow up without any experience of gardening or raising poultry unthinkable. She bought a couple of bantam hens and some fertile Araucana eggs, and soon there were nests appearing all over the barn.

One Sunday afternoon her friend Angelika brought her husband Carlo and daughters Heidi and Christina from Northampton to visit. They were all enjoying themselves so much that as the afternoon wore on, they decided to make supper for themselves so Angi and Carlo could stay longer. Omelets would be nice, they thought, so Patty started out to the barn to see if there were enough eggs.

When he found out where she was going, Carlo asked to go along. A New York City boy, Carlo had never gotten any closer to the country than visiting friends, and his friends' cottages were usually in, but not of, the country.

When he and Patty walked into the barn with Heidi, there were several Easter-colored eggs lying in the little round nests that the hens had made for themselves in the hay. Carlo was dumbfounded. "I thought they came in egg cartons!" he exclaimed.

Apparently, NRC felt much the same way about uranium.

So far as they were concerned, uranium mining was no more a concern of theirs than coal mining or oil drilling. Uranium, to them, came in neatly packaged fuel rods. There are hundreds of thousands of suffering—or dead—people, mostly poor farmers or Native Americans, who could have told them differently if they had been prepared to listen.

From Texas to Oregon, scars upon the landscape bear mute evidence of decades of uranium mining and milling operations, much of it on Indian reservations. Navajo miners developed lung cancer from breathing radioactive radon gas in the early deep uranium mines. Later, open pit miners and families living downwind from their mines were also severely affected. The children of all those who survived are suffering the effects of their parents' exposure, and their unborn children will continue to suffer for generations to come.

When the NRC talks about cheap nuclear power, they aren't adding in the cost in taxes for medical care and orphanages. They are ignoring the human price of individual suffering and of families destroyed by the lingering illnesses and deaths of parents or the despair of parents who must watch helplessly while their children die. They certainly aren't including the cost of storing and guarding the waste for hundreds of thousands of years.

After mining, rock containing uranium ore is ground into sand and treated with chemicals to separate out the uranium oxide. That process removes 15% of the radioactivity from the sand. The other 85%—from leftover uranium, thorium, radium, radon and polonium—is left in the sand. It takes 80,000 years for half of the thorium to turn into other elements, and most of the elements it turns into are themselves very radioactive.

What happens to this radioactive sand? It is stored in unsealed, uncovered ponds to dry out!

As it dries out, radioactive elements are carried into the ground water or seep into rivers and lakes. When the sand is dry, the wind is free to blow it wherever it will. Since the prevailing winds are from west to east, the eastern portions of the country may, in the long run, suffer more from this form of radioactive pollution than the West.

Near Shiprock, New Mexico, a dam on one of those ponds broke, releasing the water and sand to run into the nearby creek. Cattle drank the radioactive water and died. Indians who owned the land had the choice of killing themselves with radioactive drinking water and food grown with radioactive irrigation water or of starving to death.

Before people understood that that leftover "sand" was still very radioactive it was used in home, school, and commercial construction projects. It was used on roadways and parking lots—and in children's sand boxes. Thousands upon thousands of unsuspecting people have suffered from the effects of that radiation all their lives.

The processed uranium itself remains a terrifying problem—even before it becomes a fuel rod. The processing plants are far from the mines, and getting the uranium to the plants can be lethal. While we were waiting for the second half of the Environmental and Site Suitability hearing, thirty drums containing 21 tons of uranium oxide were flattened by an accident on a highway near Boulder, spilling the flour-like radioactive powder a foot deep in some places!

Even more frightening, however, was the aftermath: Because no one had responsibility for the safety of the shipment, no one thought to close the highway until local health officials arrived, and that was after people had been driving through the spill for twelve hours. The public was not informed until a week later, when a Boulder newspaper interviewed the NRC representatives who had just showed up and discovered what had happened. No one had noticed that background radiation had risen to 44 times its normal level.

And it took another three days for cleanup to begin because industry and government officials were too busy squabbling about who had to take responsibility.

We had been urging the NRC—unsuccessfully—to consider uranium milling and radon gas in the hearings. Then when that accident prompted Ralph Nader's Washington-based Critical Mass Energy Project to file a petition asking for stiffer requirements for emergency response plans—and when the New England Coalition on Nuclear Pollution filed another, similar one—the NRC could stall no longer. They conceded that, on an individual basis, radon222 and its health effects could be considered in nuclear plant licensing cases. Radon was also an issue of the National Environmental Policy Act and it provided CASE with an opportunity to force the opening of another hearing.

The Dollar Value of
a Human Life

For nearly three months, PSO had confidently been expecting the NRC to issue an LWA for Black Fox at any moment. Heavy construction equipment waited at the site for the go-ahead signal. From the day the Water Resources Board hearing had adjourned, crews of construction workers had been primed to start work at a moment's notice.

PSO was already very unhappy about all the delays—delays they had never expected when they began their plans for Black Fox. So it was understandable that Black Fox manager T.N. Ewing was terribly upset when he learned that the environmental and site suitability hearing would be reconvened on June 5, 1978, to consider radon releases during uranium mining and milling.

Clearing the site

Ilene Younghein in the hearing room

This latest delay "will undoubtedly increase the cost of Black Fox Station, ultimately resulting in higher costs of electric energy," he complained, making a point of disclaiming responsibility: "Events beyond our control, not directly related to the Black Fox project, have delayed issuance of the Limited Work Authorization."[1]

He concluded petulantly that, "For several months the NRC has been considering this issue. It seems more than slightly absurd that the NRC has determined that a separate hearing is now required."

In the two short days of the hearing, the NRC witnesses maintained that coal mining and burning caused as many or more cancer deaths from radon than uranium mining. To listen to them, one would have thought that through the use of nuclear reactors, they planned to save the world from radiation!

Cornell physics professor Robert Pohl dismissed that argument: "The long-term potential health effects resulting from the piles of radioactive sand completely dwarf those caused by the entire [nuclear] fuel cycle." The NRC's error was that they "considered only the first 100, 500 and 1,000 years after mining. Longer time frames can completely reverse the picture." Asking for the health impact of one nuclear plant, he explained, is like asking for the health impact of smoking one pack of cigarettes. "The crucial question is, of course, what the collective effect of all uranium mill tailings likely to be generated in the U.S. would be."

Another episode in the perennial debate on the value of human life began with a question to NRC witness R.L. Gotchy. But it was somewhat different from the usual question: "What is the dollar value of a human life?" he was asked.

And Mr. Gotchy was prepared to whip out the definitive answer! "The Department of Energy, I believe, has a planning guide out where they have proposed a value of a human life at something on the order of $200,000."

Once that particular question had been broached, it became somewhat like an unfilled cavity. People couldn't seem to keep their tongues off it. The next day, PSO lawyer Gallo asked his witness, Dr. Whipple, if he had an opinion on the question.

Whipple chose to temporize: "I think it is a very difficult matter for society to collectively put a value on the proper expenditure to save a life, even in the near term."

However, one cannot accuse Dr. Whipple of lacking a sense of humor. In a carefully rehearsed exchange, Mr. Gallo asked him what the courthouse walls were made of. "They look like marble," was the reply. "Some marble contains quite high concentrations of uranium."

"Then we may have been receiving high doses of radiation while we have been here?" he was prompted.

A ripple of laughter rewarded Whipple's reply, "That is a frightening thought."

Board member Shon was not willing to drop the subject of the value of human life so frivolously. "Just how many generations do you have to worry about? Do you worry about your great-grandchildren or your great-great grandchildren, or do you worry forever?" he asked Mr. Gotchy.

"I don't believe you have to be a lawyer to address this question!" Gotchy retorted.

We brought in epidemiologist, Dr. Stanley Ferguson, from the Colorado State Health Department to discuss the effect that the radioactive "sand" that was used in construction had had on the people of Colorado. The kinds of leukemia caused by ionizing radiation were occurring in the contaminated areas at a rate two and three times that in Colorado as a whole, and the death rate from leukemia was well above the national average.

He supported Dr. Rosalie Bertell's earlier contention when he warned that "No level of radiation, however low, is not associated with malignant change and genetic damage." But his final point was, in my opinion, the most crucial: "The mistakes were human mistakes, and not recognized by even the best minds available until quite late."

He was echoing what I am still trying to explain to everyone I talk to: We should not be deliberately putting ourselves and our most remote descendants into a situation that we cannot undo, when we can be reasonably sure it will harm us and them!

Dr. Zink announced that PSO had signed a fuel contract with Mobil Oil. The uranium was to be produced by a new *in situ* leaching process that, because it did not involve milling, had less potential for producing airborne radiation.

It used to be that after a Shakespeare tragedy, a farce was performed for comic relief. The lawyers performed a miniature farce for us at the end of the hearing. Dow Davis presented Dr. Robert Gilbert to update Jan Norris's earlier testimony.

"Excuse me, Mr. Chairman" Tom Dalton exclaimed disingenuously. "I don't have any pre-filed testimony for Dr. Gilbert. I claim surprise. I am not prepared to make an examination. I would ask that this hearing be continued to a day we can be provided with his testimony and given opportunity to examine."

Dr. Gilbert was excused from testifying.

Tom had been corresponding with the three engineers who had resigned from GE in 1976. The minute Chairman Wolfe finished announcing the health and safety prehearing conference on June 29 and adjourned the radon hearing, I went out to the lobby and phoned the engineers to ask if they could be available for the health and safety hearing. "You will have to guarantee a $10,000 deposit within a week," Greg Minor advised me.

"That's impossible," I declared, explaining that we had already been involved in three hearings that year and the last one had just adjourned. They were the only people we could think of who would serve our purpose so, fearful that he would refuse to help us, I offered him $2,000 right away with the remainder as soon as we could raise it. To my great relief, he gave his word that they would testify.

The Clamshell Sunburst

The prospect of Black Fox was not the only worry I had about nuclear power plants. At the time of all these events, two of my sisters, a brother, a niece and a nephew, lived in the near vicinity of nuclear power plants in Oregon, Nebraska, New Jersey, and Illinois. My eldest daughter and her husband and children lived in Massachusetts, five miles down river from Vermont Yankee, the problem-plagued second-oldest commercial nuclear power plant in the country, and twenty-odd miles from Rowe, where the oldest such plant is now being decommissioned. (Thanks to Sam Lovejoy, she does not have a third plant three miles away on the Montague plains.)

And that was just the beginning. Every plant in the world threatens millions of my brothers and sisters in the human family. My goal and the goal of most nuclear power oppo-

CLAM RAFFLE
am is hoping to make some mu...
money at the Fair, too!
effort is a Clam Raffl...
n on the 18th. The Grand...
M-SUN quilt. This 72″...
gned, hand sewn and qui...
f CASE, Citizens Action f...
Oklahoma. Raffle tickets are...
he Hampshire or Franklin...
y are $1.00 each; $10 for a...
will have books of tickets...
nd before, as occupiers...
gather signatures o...
BROOK petitions. We...
icycle donor for their...
s support of the...
our efforts to stop...
ngland.

Above: Carrie with her sisters, Paula Bentley, Clara Barefoot-Sehorn, & Florence Cahalen

Left: Carrie stitching the Clamshell Sunburst quilt, at Toward Tomorrow fair, Amherst, Mass.

nents was and is to stop all nuclear power plants, not just those in our own back yards!

When I had gone to Critical Mass '75, Sam Lovejoy and his Clamshell Alliance were already in the midst of a struggle to prevent Northeast Utilities from building two more reactors, this time at Seabrook, New Hampshire, a little seacoast town of 2,000 inhabitants. Many of those inhabitants made a living harvesting clams from the beaches, and the habitat of the clams was to be destroyed by the power plants' hot water discharge into the ocean. Thus the name of the group.

They held sit-ins at the site and participated in other acts of civil disobedience. Many of their members went to jail for their convictions, and some of them suffered exceedingly. When Robert and I made a flying trip to Massachusetts in the summer of 1976, I went to Greenfield early one morning, to talk to a group setting out for one of the demonstrations and wave them on their way. (Patricia tells me that when, on occasion, she sees fit to brag about "my mother, the flower child," Massachusetts people often say to her, "Oh, I remember your mother. She was there to say good-bye when I went to Seabrook!")

The *Sun in Splendor* had been such a fund-raising success, that I decided to use the same idea once or twice a year. However, in gratitude for Sam's and the Clamshell Alliance's work, I decided to make another fund-raising quilt as a gift to their cause.

When I was a child a favorite traditional quilt pattern was the clamshell quilt with its many rows of multicolored scalloped motifs. I thought it would be a clever idea to use the clamshell motif but adapt it to make a sunburst design reflecting both the name of the group and a life-giving source of energy to counter the death-dealing nuclear plant they were fighting.

At Critical Mass '75, I had met Anna Gyorgi, a member of the same Ball Farm commune where Sam lived and one of the major forces in the Clamshell Alliance, so it was natural for me to call her with my idea. She thought it was exciting: "There's going to be an alternative energy fair at UMass in June, and we will have a booth. Why don't we plan to hold the drawing there?" she suggested.

I carefully selected yellows, reds and oranges like those of the Sun in

Splendor for the central sunburst medallion and its border and placed them on a deep turquoise background to represent the ocean and its threatened life forms. The quilt was gorgeous, and the nearer the time came to send it to Massachusetts, the more I wanted to accompany it. One day I exclaimed to Robert, "I can't even think of sending this quilt in the mail!"

Robert, grinning from ear to ear, demanded, "When do we start?"

It had hardly been a year since his open heart surgery, and in many ways, he was fitter than he had been for many, many years. Every day since he had survived had become another wonderful gift. Happiness deferred could very well become happiness forever forgone, and I'm sure he was thinking how wonderful it would be to visit Patty and her family again.

Patricia was just as excited at the possibility of our spending some time with her in Massachusetts. She lives so far away and there were so many demands on our time and energy that we had rarely taken the time to visit her, expecting, rather, that she would pack up her children every year or so and come down to Oklahoma for a good visit.

The quilting was still far from being finished. We took a redeye flight, and throughout that night on the airplane, I stitched away. Patricia picked us up at the airport near Hartford, Connecticut, the day before the fair. I continued quilting all the following day, and then moved my operations to the "Toward Tomorrow" fair. Except that it was being held on the campus of Massachusetts' land grant university at Amherst, and many of the exhibits were aimed at a more severe climate, it was very like the Energy Expos that Ron Surface put together.

The first day of the fair, Anna Gyorgi set up a table and chairs for me in a geodesic dome exhibit. As usual, I was stitching away as I talked to the people who came in, explaining the dangers of nuclear power and the potentials of renewable power resources. People would stop to see what I was doing, become engaged in conversation, and usually take a ticket. Two of my visitors were totally unexpected: John and Linda Jacobs-Danner—who were frequent CASE meeting hosts—had, unknown to me, also decided to go to the fair. Their principal reason for being there was to attend a workshop on civil disobedience.

Another visitor was a retired Amherst librarian named Winifred Sayer. She came into the dome dressed in hiking clothes and wearing a knapsack. After she filled in her ticket, she sat down to visit with me for a while. I should probably say she sat down to educate me, while I continued the quilting on the Clamshell Sunburst.

It was from Winifred that I learned that there are usually two organizations in a given area, each working in its own way to stop nuclear power. I had not realized that the Clamshell Alliance was not pursuing legal interventions against the Seabrook nuclear reactors. Instead, they focused on civil disobedience while the New England Coalition on Nuclear Pollution, along with a number of others including the New Hampshire Audubon Society, were the legal intervenors. My conversation with her gave me new insights into some recent events in the struggle in Oklahoma.

My brother Marvin and his wife, Rita, had come up from New Jersey to visit their daughter in Boston, and all three of them came out to spend a day. My niece Sandra spent that day at the "Toward Tomorrow" fair with me and helped solicit contributions for the quilt. She was a marvel. At the end of the day, she decided to stay and help me until the fair was over, and I was very grateful. Those two days, instead of staying put and waiting for people to come to me, both of us roamed over the whole fair, meeting people in booths and collecting information about the exhibits. In all, she helped me collect nearly $2,500 for the Alliance.

On the last day, a little girl who had been dancing all over the fair for the three days drew the winning ticket. (I have often wondered if that child were the daughter for whose sake Sam had downed the tower.) When we read the name of the winner, the new owner of the quilt was the same Winifred Sayer who had taken the time to share with me what she knew about the politics of intervention!

After the drawing, Anna Gyorgi asked Winifred if she would allow the Alliance to take the *Clamshell Sunburst* quilt with them to Seabrook the following day so they could use it as a banner for the sit-in they planned. Anna promised to take good care of the quilt, and of course I, too, agreed. The sit-in was featured in many news photographs and television news spots, and in each one, the *Clamshell Sunburst* waved proudly over the heads of the protesters.

The weather was perfect those three days at the fair—and equally perfect in Northfield. It inspired Robert, who had never in his life been able to see things that needed to be done without itching to get started. What he saw was the sagging remains of an old shed attached to his daughter's house that needed to be dealt with. Before they knew it, he had gone to the hardware store and bought a chain saw to remove the part that needed to be replaced and climbed up on the roof to cut it. Next, he dug a two-foot trench for a new footing, arranged for the local sand and gravel company to deliver the wet cement, built the above-ground forms, and supervised the pouring—all this in the first two days.

The next day he devoted to buying the materials for the remainder of the repairs, arranging for and supervising their delivery, and before leaving, setting down careful instructions for completing the project. Many perfectly able-bodied men would not have been able to accomplish what he did in a week, and he had been at death's door only a year earlier!

Despite all the civil disobedience by the Clamshell Alliance and the years of intervention by the New England Coalition, despite all the legal barriers put in its way by then-Governor of Massachusetts Michael Dukakis, the NRC finally prevailed in Seabrook—but only partly: Only one of the reactors was built.

Raffles, Rallies, and Rock 'n Roll

One day, when I was buying fabrics for the Clamshell Sunburst, I went into the Handmaker's Shop in Tulsa. As I compared and matched colors and patterns, a man came in the door carrying an antique trunk. "Could you take a look at these and tell me what you think?" he asked the owner, as he set it down and opened the lid.

She bent down and pulled out a beautiful floral-appliquéd quilt. The two dozen or so people in the shop started gathering around the two of them as she said, "I think this should bring about $500." For some time, she kept bringing out quilt after beautiful quilt and estimating their values at several hundred dollars. Knowing all of them were too expensive for my pocketbook, I was about to turn away when from the bottom she pulled out two Double Wedding Ring quilt tops that had been pieced but never quilted. "These should go for about $30 apiece," she declared.

"I'll take them," I put in quickly before anyone else had a chance to open her mouth. It was one of the best bargains I ever made.

It turned out that the man had bought a house in Tulsa that had

Photos © Dan Agent

changed hands a number of times. As he cleaned out the debris in the attic, he came upon a stack of rubbish. At the bottom of that stack sat the old trunk full of treasures that had probably been made in the thirties. The fact that for perhaps 40 years they had escaped moths and mice was a minor miracle.

A Claremore woman, Lessie Wetherell, had been making quilts for other people most of her life. I bought about $20 worth of batting and lining materials for each quilt and took the whole thing to Mrs. Wetherell. She did herself proud, quilting the background areas in a simple, but effective design that echoed the contours of the pieces, a design that I had never seen before. She had said she would only charge me $30 for quilting each quilt, but I paid her $100 apiece and felt like a thief when I could not afford to offer her more.

Nervous as I was about getting in trouble with the law, I never sold lottery tickets. Instead, I would accept a donation of whatever amount the person chose to offer and make out as many tickets in that person's name as there were dollars in the donation. For a person who had been shy and retiring all her life, I was shameless about asking everyone I saw for a contribution. And once Mrs. Wetherell had finished quilting the double wedding ring, I started carrying it everywhere with me to show to people.

One day I went into a Tulsa health foods store to replenish some of my vitamin supply. The owner was a CASE sympathizer, so when I had

Nuclear Rally-Concert

Gregory and author Kurt Vonnegut Jr.

Nix said Browne promised to bring three other "nationally known singers," each of whom sang during the Washington rally.

Included among the Washington participants were Bonnie Raitt, Jesse Colin Young and Buffy Ste. Marie. They have expressed interest in the rally but have signed no contracts, promoters said.

Browne told promoters he also would bring David Lindley, a member of his band, who accompanied Browne here in September.

Proceeds will be split four ways, Nix said, with funds going to Sup-porters of Karen Silkwood Citizens'

in South Carolina.

Barnwell Alliance is included because that is the site where the nation's first nuclear waste storage facility is proposed, Nix said.

The rally-concert will be on the Mohawk Park polo field, about one mile east of the zoo. The field is about 20 acres, according to Tulsa Parks and Recreation Director Hugh McKnight, who said there are no residences nearby.

Quake Hits Mexico

MEXICO CITY (AP) — A moderate earthquake struck southern Mexico near the Guatemalan border

taken care of my purchases, I asked him for a contribution. "Carrie, I just don't have the extra money right now," he responded.

As I put my packages in the car, my eyes fell on the quilt, and on impulse I decided to take it back into the store and see if either of the sales ladies wanted to make a contribution. The owner heard us talking and came out of his office. As soon as his glance lit on the quilt, he exclaimed, "I'll take ten chances. No, make that twenty!" and he pulled a twenty dollar bill out of his shirt pocket.

That episode taught me a lesson: People are more willing to contribute when there is the chance of their getting something material back.

We planned to give away the double wedding-ring quilt at a rally to be held a week after Robert and I returned from Massachusetts. One woman had sent a $100 contribution. She hadn't known about the quilt when she sent it, but she was just as much a contributor as anyone else, so Robert helped me write her name on 100 of the tickets.

Suddenly he said, "I should have my name on some of those tickets. I've contributed more than all the others combined."

Robert's Grandmother Hanes had thirteen children, and she had to keep them all warm at night, so when she sat down in the evening after a day's work, she would always pick up her latest quilt and set stitches until it was time for bed. When Robert's parents were married, two of the gifts they received were a *Grandmother's Flower Garden* quilt with its thousands of minuscule hexagons and a *Double Wedding Ring* quilt. There was rarely a time in their married life when their bed was not covered by one or the other.

I could see a lifetime of memories, happy and sad, in Robert's face when he looked at the quilt, so I wrote his name on 200 more of the tickets. (I knew my hand wouldn't hold out long enough to write out one for each of the several thousand dollars he had contributed that spring to pay attorney and expert witness fees.)

At the Riverside Park rally, Janie Duenner added a sparkle to the day as she sang to the accompaniment of the guitars of Michael Long, David Abrahamson, and Marilyn Oldefest and Phyllis Ellias' autoharp and dulcimer. (That dulcimer performance was a first for me, and I thought I had never heard anything so beautiful.)

By the day of the rally we had raised $3000, not counting Robert's contributions. Robert was involved in other things on the day of the rally, so I went without him. Much to our surprise, he won the quilt. He had felt perfectly comfortable taking tickets, but he was embarrassed to have won.

"I can't keep it, Carrie," he said. "Why don't you raffle it again?" This time, we raised $5000. All told, my $150 investment and beautiful work of Lessie and the unknown quilt-maker brought CASE $8000!

The second drawing took place at the Energy Expo that fall. The five thousand tickets were folded so they wouldn't cling together or take up too much space. Even so, we had no container large enough to mix them in, so we dumped them on the floor. They looked like a pile of autumn leaves as we turned and tumbled them. The gentleman with the renewable energy truck—and its little house built on the back—kindly agreed to draw the winning ticket.

Closing his eyes, he reached clear to the bottom of the pile. The name on the ticket he fished up was Cathy Coulson-Currin!

Cathy had never before owned a quilt. Her four children were all to sleep under it as they were growing up. After the Black Fox cancellation Cathy brought it back to me, asking me to replace some of its pieces that had become frayed and worn. The opportunity to do so never seemed to materialize, and when Cathy eventually moved to Oregon, I suggested she try to find someone out there to do the job.

"It's yours, Carrie," she told me. "Robert won it fair and square the first time, and my children and I enjoyed it for years. You keep it." Today, it hangs on the wall of my living room, delighting the eyes of all who visit.

The *Sun in Splendor* and the *Double Wedding Ring* were only the first in a succession of quilts that, with a host of other donated items, helped us raise over $60,000.

Because of the prevailing wind patterns, Northeastern Arkansas would have been affected by a nuclear accident or meltdown at Black Fox, so I packed up my films and took them to meetings in Fayetteville on several occasions. Through those meetings, People's Action for a Safe

Environment (PASE) was organized. Also, because of those meetings I met Dr. Benjamin Spock and his wife, Mary Morgan, who lived in a solar home at Rogers, Arkansas.

In 1979 Barbara Jordon and twelve other PASE women decided to make a large quilt for CASE. Their original and striking *Dutch Windmill* design was of sixteen different Dutch windmills on sixteen blocks. On its large center block, the wind blew through a larger mill. Their quilt drew all eyes as it hung behind us in the CASE/Sierra Club booth at the Tulsa State Fair and, later, at the Energy Expo that fall.

The mother of another longtime Arkansas friend, Carmyn Pitts, had pieced the blocks for a *Rocky Road to Kansas* quilt, but never completed its assembly. Carmyn gave the blocks to Jessie Fink, and she and her mother put them together and quilted the finished top for another drawing. Then Jessie and her mother made a *Log Cabin* quilt for us to give away.

Soon everybody wanted to get into the act. Karen Collins, from Tacora Hills near Oologah, made a *Roman Stripe* quilt. Philbena Janz from Inola made two baby quilts, and Karen Economides of Jenks pieced a *Bear's Paw* baby quilt and gave it to me to finish. (Joe Clement won that quilt for a new grandson.) I myself appliquéd twelve copies of my quilted *Oklahoma Horizons* wall picture with its scene of Oklahoma hills and dales embellished with an appliquéd sun peeking from behind an embroidered windmill. Joan Breit and Nadine Holly helped me embroider the windmills. (Of the twelve, I know Joe Clement bought one and Eva Unterman's daughter won another, but I have been unable to find my records of the other winners.)

Over the years the drawings included items of every description, including a bicycle and a motorcycle. On several occasions Win and Kay Ingersoll contributed one of their organically raised cattle that had been butchered and frozen. (On four occasions, the beef winners were vegetarians, so we had to draw again to find someone who could properly appreciate the largesse of the Ingersoll Ranch.)

The money from the Riverside Park drawing was very welcome but far from sufficient to cover CASE's commitments. I discovered, through some of my friends in the anti-nuclear movement, that for some time the

California-based Pacific Alliance had been sponsoring anti-nuclear fundraising concerts by well-known musicians up and down the eastern and western seaboards. I wrote to John FitzRandolph of the Alliance, asking him if they could help CASE.

At the same time, Karen Silkwood's family was suing Kerr McGee over her death, and they, too, needed money for legal fees. A month after I wrote, I received a letter from John, telling me that Jackson Browne would be giving a concert at the Tulsa Performing Arts Center in the fall to benefit CASE and the Silkwood family!

Jackson Browne's sold-out concert was on September 3, 1978. He was a delightful young man, and I loved his music. So did the hundreds of people who came from all over Arkansas, Kansas, Missouri, and Oklahoma to hear him. It was one more opportunity to spread the word about nuclear power and, at the same time, to take some time to enjoy ourselves.

Not least in my enjoyment was the $2000 that turned out to be our part of the proceeds.

Jesse Colin Young & Jackson Browne

©Dan Agent

TULSA SUN FESTIVAL
FOR A NON-NUCLEAR FUTURE
JACKSON BROWNE, BONNIE RAITT,
JESSE COLIN YOUNG, BONNIE O'KEEFE
MOHAWK PARK TULSA, OKLAHOMA
DAY JULY 8th –2:00 P.M.
NO CANS OR BOTTLES
No. 7806

ERAL ADMISSIC
No. 7806

Carrie with Bonnie Raitt

Thirty - Two

Sunbelt Alliance

By 1978, there were three anti-nuclear groups working diligently to defeat Black Fox. They were CASE; John Zelnick's group, Citizens Against Black Fox; and a group of Tulsa University students organized by a young man named Stephen Henry, who had composed the clever acronym CARE (Citizens Against Radioactive Exposure) for their name. Stephen worked tirelessly against nuclear power. He did an enormous amount of research, and then printed his findings and dispensed the information to thousands of people. When he died in a tragic automobile accident in April, 1982, we lost a wonderful friend as well as a devoted worker against nuclear reactors.

CASE had been holding our meetings in the home of John and Linda Jacobs-Danner, in north Tulsa, for a number of months. The Jacobs-Danners had come to

Pr

Me
said
date
the s
nucl
Inola

the Sunbelt Alliance they will announce a y to begin occupying proposed Black Fox erating plant near

The alliance is protesting the project, planned by Public Service Co. of Oklahoma, and has ~~ht it in earlier legal proceedings PSC spokesman Joe fied of the alliance's ann

commented, "Our only contingency plan is to love 'em to death."

An alliance spokesman, Jim Garrison, said members will move onto ~~d begin planting trees and ~~re the site to its ~~aid pro-

believe that we would be more effective if the three groups were to work more closely together. We could cooperate in raising funds, and people who belonged to more than one organization would not have to attend so many meetings. It seemed reasonable to the rest of us, and, accordingly, we all joined together under an umbrella organization called the Sunbelt Alliance.

"Hallowe'en Thirteen" at Black Fox site

"Alliance" was a word that had been appearing more and more in the names of anti-nuclear groups. While those groups seemed to be more involved in civil disobedience than in legal intervention, those of my generation — in Oklahoma — saw that as being mostly a natural concomitant of their having been organized in places like California and Massachusetts. After all, it was on the Berkeley campus at the University of California that the teach-ins of the Vietnam War era got their real start — not Tulsa University — and it was Harvard that came to our minds when we thought of students occupying administrative offices — not Oklahoma State. We lived in Oklahoma, after all, and young people from Oklahoma were sensible enough to know that civil disobedience would not convince anyone here of anything — except for people their own age.

Although our intervention had delayed it for many months, PSO's

Limited Work Authorization was granted on July 26, 1978, a month after the Toward Tomorrow fair. Tom Dalton immediately appealed to a higher court, but that didn't stop PSO. In fact, they were so anxious to make a point of their victory that they made a kind of carnival of it—an exclusive carnival designed to give the impression that the opposition didn't exist.

They invited so many dignitaries from all over the state and region to their groundbreaking ceremony on August 14, 1978, that bleachers had to be constructed to seat the crowd. Among those who were *not* invited were Carrie Dickerson, Ilene Younghein, John Zelnick and the Jacobs-Danners. The closest I came to the ceremony was watching Governor Bellmon on television playing at operating one of the big ground-moving machines.

Members of the Sunbelt Alliance, however, crashed the party and mounted a protest. PSO had arranged for Rogers County deputy sheriffs to handle security for the ceremony. Normally that would not have been a problem, because Rogers County Sheriff Amos Ward was very professional in the supervision of his department. Unfortunately, one of the deputies must have taken his job too seriously. Jim Sellman, a photographer who participated in both CASE and Sunbelt activities, was severely beaten as he took photographs of the protest and was taken to the hospital. Jim's martyrdom really set things buzzing.

The Sunbelt Alliance met frequently that summer, and the leaders started making plans for a Black Fox occupation the week-end before the health and safety hearing was to convene. I believed that the occupation was much too premature. I told the Alliance leaders that if all else failed I would be right there with them, but I believed we would win through the Oklahoma Corporation Commission rate hearing — not the NRC hearings. "Oklahoma is a very conservative state," I told them, "and if you follow through on the occupation, we will all lose credibility with very many people."

Nothing I could say moved them, and finally the CASE board decided to withdraw our organization from the Alliance rather than participate in civil disobedience. CABF and CARE followed suit. Many of our members, however, chose to join Sunbelt as individuals, and most of the younger ones resigned from CASE. Fortunately, most of those who wished to continue with civil disobedience understood that we were not rejecting them or our mutual goal, only differing in our convictions about the most productive way of proceeding. It was personally devastating for me, however, that some of the Sunbelt members took the attitude that "If you are not with us, you are against us." Even worse, this stance caused a discord we could ill afford if we were to be as effective as possible. Eventually, most of the people who left CASE for the Sunbelt returned, but for many months those of us who were left fought a very lonely battle.

Young people who had participated in other direct action campaigns on the east coast came to Oklahoma to help organize the "occupation." They went from campus to campus, organizing college students at OSU, OU, NSU and TU. Their favorite statement, "We cannot stop Black Fox through the legal route," was very persuasive to young people who had spent most of their lives in passive roles in classrooms. When they heard, "Climbing over the fence and getting arrested at the Black Fox site is the only way to stop Black Fox. Join us and we'll stop the nuclear power plants," they felt that finally their energy and passion could go into something productive and worthwhile.

Scores of students and most of the CASE members joined in their activities. I soon realized that nothing would stop the occupation and that many people were being hurt by divided loyalties, so I finally told everyone that if they had to go ahead with it, they should make it a tremendous success. "Everyone but Carrie Dickerson should be there," I said.

That cold, drizzly Friday night, approximately four hundred people from all over the country camped out at Rocky Point Recreation Park. The leaders had done careful planning, organizing affinity groups and setting up support groups to take care of belongings that were to be left behind in the park. Portable stiles had been constructed to make it easier to get over the fences without causing damage. The long line of protesters sang as they trudged through the rain and mud about a mile into the off-limits area. Those who had children with them, carried the little ones on their shoulders. Amos Ward and his deputies were waiting to arrest them and load them into buses. They were then carted off to the Rogers County Courthouse in Claremore for arraignment.

Then on Halloween night, during a recess in the health and safety hearing, fourteen members of the Alliance (the Halloween Fourteen) managed to elude the vigilance of the PSO guards as they stole onto the grounds and chained themselves to bulldozers and other heavy equipment.

There is nothing reporters like better than a spectacular "media event," and both of these acts of civil disobedience qualified. Media representatives were there in droves, and a number of reporters were arrested along with Sunbelt members. The attention of people in the state focused on civil disobedience to such an extent that it almost eclipsed reports of the health and safety hearing.

Granted, all this attracted lots of attention and made more people aware that fellow citizens were deeply concerned about the dangers of nuclear power. Yet in this conservative state of Oklahoma, it had an even greater negative impact. When I was invited to speak to service clubs like the Lions and Rotary, the first question was always, "What have you gained by crawling over the fence and getting arrested?"

It took so much of my time just to explain that CASE had not participated in the occupation, that it ate into the time I had to discuss the economic foolishness and safety threat of the nuclear monster and the advantages of other energy solutions. Finally, I designed a logo, a windmill and sun disk circled by the words, "Citizens' Action for Safe Energy ... the LEGAL INTERVENOR against Black Fox," and had it printed on the back of my jacket and on aprons and T-shirts.

The Sunbelt Alliance's civil disobedience events had an immediate and devastating effect on CASE's ability to raise money for the legal intervention. Not only did the media give Sunbelt extensive coverage, few of the reporters even understood the important distinction between civil disobedience activities and legal intervention, and even when they did understand and wanted to make it clear, it was difficult for them to do so in sound bites and headlines—which were what most people paid attention to.

This had a double impact. First, people who were attracted to civil disobedience sent their money to Sunbelt—either instead of CASE or thinking that they were also contributing to CASE. Even today, those people see me on the street and stop to tell me they sent contributions to the Sunbelt. When I look at their faces, I see that they are clearly expecting me to thank them for their support in my work. By now, it's all water over the dam, so I thank them kindly, and tell them how much I appreciate their efforts.

Equally disastrous, many of our—formerly—most reliable supporters said they could not in good conscience continue their support. I spent inordinate amounts of time, first telling them, then trying to show them that CASE was not part of the civil disobedience. True we had the same goal, to stop Black Fox, but our methods were distinct and our needs were different.

After Three Mile Island, however, Sunbelt raised about $200 for CASE and they cooperated with us in bringing Jackson Browne and David Lindley back to Tulsa along with Bonnie Raitt, Jesse Colin Young, Danny O'Keefe and Freebo (from Bonnie Raitt's band) for a July 8, 1979, anti-nuclear concert. These wonderful entertainers were joined by local artists Randy Crouch, Michael Long, Marilyn Oldefest and the Blue Cliff Ensemble. John Trudell, the National Chairman of the American Indian Movement, and Sara Nelson, from Supporters of Silkwood, had been invited to speak.

We expected six thousand people, but eight thousand had shown up in Mohawk Park by mid-afternoon for the concert. An elaborate stage had been built, and the crowd sat on quilts and blankets. We set up booths under the trees bordering the area to the west. CASE set up tables for our T-shirts and information sheets. Karen Long-Economides, Fenton Rood and David Martinez organized the group that sold hot dogs and drinks to raise money for CASE.

The Pacific Alliance had again sponsored the concert, sending Tom Campbell to coordinate things with the help of a local attorney, Jeff Nix. The proceeds after expenses were divided among four groups, CASE, Sunbelt, Supporters of Silkwood, and the South Carolina Palmetto Alliance. Our portion of the proceeds amounted to about $2000.

Most of the musicians giving concerts for anti-nuclear groups believed that legal intervention was useless and that only civil disobedience was worthwhile, and at the time of the concert, most CASE members were more mature people whose preferences in music were quite different from those of the younger members. Many of them stayed home. I participated in the committee meetings where we prepared for the concert, but CASE was not given any responsibility in the actual physical preparations of setting up the stage and the fences and gates behind the stage. Those activities were in the hands of the younger, stronger people, and most of our young members had defected to Sunbelt. Security and traffic control was in the hands of Sunbelt members, and the people in charge turned me back when I attempted to go backstage.

However, one of my friends in Sunbelt came to the CASE booth to tell me that Jackson Browne was planning to speak backstage to Sunbelt leaders and that I should be there. Again, the guard at the backstage gate, a young man from Oklahoma City, refused to allow me in, but this time he said, "You are not welcome here."

Shocked, I told him firmly, "I have a right to be here!" I pulled on the gate, and he tugged it in the opposite direction. Finally, with all my strength I pulled the gate open and entered, ending an undignified—and perplexing—episode.

When I arrived backstage, I found a number of Sunbelt members sitting with musicians around a table listening to Jackson speak. After his speech, I thanked him and the other musicians, including Bonnie Raitt, for their generosity in giving their time and talents. Taking him aside, I then asked Jackson if he could arrange five minutes for me to tell the crowd about the history of our fight to stop Black Fox and to thank them for their support. He very kindly helped me onto the stage and introduced me.

As I left the stage I saw Tom Campbell, so I went up to him to thank him for bringing the concert to Tulsa. In an angry voice, he replied, "CASE has no right to a penny of the money from this concert." Astonished at this rebuff I staggered off-stage, wondering what I had done to provoke it.

CASE's debts always accumulated faster than I could find the money to pay them, but even so, it took two years before I got so desperate that

I once again phoned the Pacific Alliance, dreading to hear that same anger in Tom Campbell's voice. He answered civilly enough, so I gathered my courage to tell him that it looked as if we could win the fight if we could raise enough money to continue through the OCC hearing. "Could you possibly ask one of the Alliance musicians to come to Tulsa and give another benefit for us?" I pleaded.

Abruptly he demanded, "Are you holding it against me for what I said to you at the Jackson Brown concert?"

"I still don't know why you said CASE didn't deserve a cent of the money from that concert. But I'm not holding it against you," I assured him.

"If you don't know," he interrupted, "then forget it. It's a thing of the past. We'll start today with a clean slate. Write a letter telling us what you need, and we'll see what we can do."

I still wonder what it was all about. It must have been something serious, or he wouldn't have remembered it two years later, but I suppose it will remain one of the unsolved mysteries in my life. Whatever it was, Tom came through handsomely for us, arranging for Bonnie Raitt to give a concert at Cain's Ballroom on March 27. She brought with her Jimmy Byfield and the Brothers of the Night, and they and a group of local musicians recruited by Tom Bomer, performed to a standing-room-only audience.

Tom Bomer, who acted as our liaison with Tom Campbell and the Pacific Alliance, succeeded in working out an understanding that all the money from the concert—which turned out to be almost $9000—would go to CASE and would not be divided with any other group. That money paid some of our debt and allowed us to continue the intervention for a few more months.

The day after Bonnie Raitt's concert marked the second anniversary of the Three Mile Island-2 accident, and groups across the country were remembering the date with various activities. CASE decided to hold a press conference that morning and to show a videotape of TMI-2. Afterward, we marched downtown to Bartlett Square, where we held a rally to publicize safe alternatives to nuclear power.

Bonnie, who had recently moved to Tulsa and so was especially concerned about Black Fox, told the crowd, "The same mentality that puts poisonous preservatives in our foods so they will last longer on the shelf also puts poisons in the environment. ... If we can control our energy needs, then maybe we'll get some control over who's telling us lies and who isn't."

It was always a pleasure for Robert and me when Bonnie Raitt came to town. She was a delightful young woman, raised as a Quaker and knowledgeable from an early age about disarmament issues and nuclear power. She dedicated much of her time and energy to political change, farm workers' rights, and the anti-nuclear movement. She was also—in a roundabout way—a reminder of Robert's mother and her family.

When Robert's mother, Anna Hanes, was a young woman, she taught school for a number of years, and one of her students was a boy named Lynn Riggs. Lynn Riggs later wrote the play *Green Grow the Lilacs* from which came the musical *Oklahoma!* In his play he used the ranch of Mr. Skidmore as the setting for a pie supper where the hero, Curly, and the villain, Jud, confront each other. Mr. Skidmore was a real person, related by marriage to Anna's family.

Before she met Robert's father, Anna's steady beau was a young man named Manuel. Anna was extremely fond of him and considered him a young man of character, but she felt that he was handicapped by his upbringing in an extremely well-to-do family. He was not "serious" enough. And one of the proofs of his lack of seriousness was the price he paid for her pie at a pie-supper fund-raiser—$100! That was a small fortune in those early twentieth century days—enough money to furnish a house. Could such a man be depended upon to make proper provision for a family? Clearly not.

However, the story of Manuel's contribution was soon all over the district, and Lynn Riggs was one of the people who heard it. Robert's mother's famous pie became Laurie's famous pie, auctioned off at a pie supper to build a school house in *Oklahoma!*

What does any of this have to do with Bonnie Raitt? Her actor/singer father, John Raitt, played Laurie's beau Curly, who bid against Jud for the pie in the first broadway production of *Oklahoma!*

Parting the Red Sea

The two-day prehearing conference on June 26, 1978, seemed cut and dried. Tom Dalton, of course, questioned the logic of going on to the final hearing before the water issues had been decided, but since the OWRB had issued their permit, that was pretty much *pro forma.*

On the second, and final, day of the prehearing conference, the *Tulsa Tribune* softened up enough to reprint a UPI article reporting that electric utilities operating nuclear power plants were granted rate increases that averaged 27%

Above:
Joe Farris

Left: Sunbelt Alliance members lined up in hearing room

shouldn't get the per-
mit to allow construc-

ek

r moratoriu

**Most opponents concede the con-
struction permit probably will be
granted next year but few are willing to concede that the plant will
ever operate.**

"Right now, I'd say 50-50," says nuclear opponent and attorney Andrew
Dalton of Tulsa.

Try 100 percent, respond PSO

which need to be made to Black Fox three-
first before a construction permit per-
asked is issued and safety-related p'
m on work, such as the reactor base
lifted, mat, begins," Risk said.
Before the NRC clamped a

federal courthouse.
Most opponents

tion of the estimated

more than those given utilities with conventional power generators. That was pretty funny, since the utilities had for years been hammering away at the idea that the risks of nuclear power were worth it to save consumers money, and the *Tribune* had for years been slavishly agreeing with them.

After the conference, Tom told us that he was so swamped with the LWA, NPDES, and water allocation permit appeals that he wouldn't have time to prepare for the health and safety hearing. Where, I wondered, could we find another attorney as willing to stick his neck out as Tom Dalton? Again, Kathy Groshong came to the rescue and persuaded Joe Farris of the Tulsa firm Green, Hall, Feldman and Woodard to represent CASE. His colleagues John R. Woodard III, Maury Efros, and Nancy Wood assisted him.

As usual before every hearing, the CASE board gathered to talk about where we would get the money to pay attorneys' and expert witnesses' fees. Joyce Nipper, Marjorie Spees and Helen Geary, like numerous other CASE members, were always ready to drop what they were doing and come to my aid. They hosted this meeting, held at Helen's home. I asked if any of the board members had friends who had money and might be persuaded to help us. Joyce Nipper promised to talk to a friend with vast holdings in oil, who contributed regularly to political campaigns. After she had read all the reports and fliers Joyce gave her, the lady called to say that she would contribute $500 to our cause and that I could pick up the check that evening.

As I drove to her home I prayed, "God, please put it in their hearts to write that check for $5000 instead of $500." When I looked in the envelope she handed me, I saw that indeed it did contain a check for $5000!

I was speechless. My prayer had been granted. I wanted to hug her, but since we were complete strangers, I was afraid that she might be offended, so I contented myself with giving both her and her husband a firm handshake and a heartfelt "Thank you!"

Joyce's friend came to several sessions of the hearing and was so impressed by the way our attorneys and expert witnesses conducted themselves that she volunteered to raise $20,000 of the money we needed for their fees. Later, she was to renege on her offer, but for weeks I was able to live in a fool's paradise.

After the prehearing conference, the licensing board notified us that they would not accept several of our contentions, questions that we wanted the public to hear: Could PSO afford to build the plants? How could they maintain that the plant's spent-fuel pools would be safe? Would the plant design prevent sabotage?

Dauntless Ilene struck back in an interview in the *Tulsa World:* "The licensing board has done an ultimate act of tyranny in throwing out our serious contentions," she asserted. The situation looked bleak indeed, and many CASE members were ready to throw up their hands and quit.

I knew our only hope was to string out the hearing process until PSO applied for a Construction-Work-In-Progress rate hike for work done under the LWA. I believed we had a significant chance of stopping Black Fox at the Oklahoma Corporation Commission hearing. There we would prove Black Fox would be so expensive to build that consumers wouldn't be able to pay their electric bills.

The weekend before the hearing began, when the Sunbelt Alliance was preparing to occupy the Black Fox site, I sent a large poster out to the gathering. At the top was printed in bold letters:

> The NRC has denied a number of the CASE contentions (issues) we had planned to have discussed in the hearing. If you sincerely wish to stop Black Fox please sign here that you wish to make a five-minute limited appearance statement. These will be heard on the morning and evening of October 10 and the evening of October 11 at the Camelot Inn Annex at Highway I-44 and South Peoria. Please tell the NRC board how you feel about nuclear power and Black Fox. Please instruct them that you wish them to reinstate our contentions, and please emphasize the fact that you wish the NRC to allow the secret GE Reed Report to be addressed in the hearing.

When the hearing began on Tuesday morning the poster had been returned to me with 400 signatures!

Running through this epic like boldly colored yarns in a tapestry are instances and events that, together, eventually helped make possible our victory against Black Fox. The story had already been unfolding for six years before the colors of the strand that represented the limited appearance statements could be seen clearly.

The Health and Safety Hearing[1] was scheduled to go from October 10 to October 20 and to be followed shortly thereafter by a construction permit for Black Fox. However, when the hearing opened with time set aside for limited appearance statements and the room was filled to overflowing, it didn't take tremendous insight to see that it was going to be difficult for the hearing officials to adhere to their schedule.

In earlier hearings, the limited appearances had very little effect on the committees' deliberations aside from delaying the procedures. This hearing was different. Before the statements began, Chairman Wolfe cautioned us that, "The limited appearance statements do not constitute evidence. If the statement of a limited appearance individual raises an important issue, a health and safety issue, this will alert the Board that there is such an issue and the Board will proceed to direct the parties to present evidence on that issue or issues."

CASE members, of course, could always be depended upon to show up in large numbers when we needed people to make limited appearance statements. This time, however, a host of Sunbelt Alliance members also rallied to the cause. In many cases people belonged to both groups, but the younger people who mostly constituted the Sunbelt Alliance often had more freedom to drop everything and attend. I believe that to a large degree it was their statements that turned the tide in the case of the Reed Report and the delays caused by their additional numbers gave us the extra time we needed.

There were two limited appearance statements in favor of Black Fox, one from the executive director of Municipal Electric Systems of Oklahoma and the other from a General Electric employee from Edmond—who assured us that his opinions had nothing to do with his job.

Many of the statements against Black Fox were simply reiterations of sentiments that CASE members had been trying to get the NRC to listen to for years. Seen from a perspective of fifteen years, some of the episodes take on a holiday atmosphere, but at the time the people involved were deadly serious: Cathy Coulson-Currin brought her son to testify, and an Arkansas schoolteacher brought a carload of her pupils. None of the children were allowed to speak. Chairman Wolfe read Dr. Spock's telephoned statement.

In a wry foreshadowing of an editorial two years later, Doris Gunn held that, "We didn't freeze in the dark before the nuclear industry took over and I don't think we'll freeze in the dark without it."

CASE secretary Eddie Bryant and Sunbelt Alliance co-founder Linda Jacobs-Danner led off a long roster of people who pleaded and demanded that GE's Reed Report be released. Many people felt that PSO's choice of a GE reactor (like the one at Browns Ferry) was dictated by cost-cutting considerations, one quoted an NRC document that declared, "GE reactors leaked a little more than twice as much radiation into the air as the Westinghouse reactors, which are a little more expensive than the GE reactors." Another mentioned NRC Chairman Joseph Hendrie's justification for not dealing with the issue because it "could well be the end of nuclear power."

A number of people talked passionately about moral issues and about the place of government in a democracy. The Sunbelt's Jim Garrison could not resist a little grandstanding. After paraphrasing Thomas Jefferson that when the government refuses to listen, we have the right and the obligation to take our message to the village square, he turned to the audience and implored, "Everyone in this room who's ready to take it to the village square, stand up."

Everyone in the audience stood, and Jim announced dramatically to the Board, "We are witnessing right now the beginning of the end of nuclear power in this state!"

Hal Rankin gave the board fair warning of the Sunbelt Alliance's stand on this issue when he quoted the Mahatma Ghandi: "In those instances where democratic procedures have been damaged through default or design, and where the legal machinery has been turned toward a travesty of justice, civil disobedience may be called into play."

As CASE chairperson I was represented by counsel, so I was ordinarily not allowed to speak at the hearings. However, the board gave me special permission to speak at the Saturday session, and I expressed my gratitude. I did not say anything they had not already heard, but it made me feel better to be able to tell them myself rather than through an intermediary.

A number of people had not yet been heard when the board halted the limited appearance statements promptly at noon on Saturday. "There are people who still want to speak," cried Jim Garrison, and he and several dozen Sunbelt Alliance members lay down side by side on the floor, blocking the way to the door.

The board members appeared perplexed as they stood at the edge of the mat of prone bodies. The young people lying on the floor were still at the age when we all think we are immortal, but as a person of about the same age as the board members, I knew how I would have felt had I been facing that barrier.

What if they try to make their way through and fall down and break something? I asked myself.

But you can't do anything! my craven self replied. *They won't listen to you.*

Oh yes I can! I told it firmly.

"Let the men pass," I said quietly, but with all the authority of a Charlton Heston in *The Ten Commandments*. To my relief and astonishment, the demonstrators parted like the Red Sea, scooting to left and right.

Quickly grabbing the mike, I gave the demonstrators something else to think about: "To stop nuclear power, the NRC law must be changed. There is a nuclear moratorium bill in Congress, authored by Alaska Senator Mike Gravel. We are trying to get a million signatures on a petition to get that bill passed by Congress. Will each of you please take with you a sheet of blank petitions and get them signed and return them to me so I can send them to Washington? Please help us get that bill passed."

Then I thanked everyone who had testified, expressing special appreciation to those who had asked for the GE Reed Report, and we went home.

David vs. Goliath

Dow Davis' opening statement, on behalf of the NRC, summarized all the changes PSO had made in the plant design in the three years since they had applied for a construction permit. "The plant as it is currently designed can safely be constructed and operated," he asserted, although "certain safety features about which the details are not known now can safely be left for resolution later on in the licensing process."

His former supervisor, Joseph Gallo from PSO, chimed in like Tweedledee. "We ... believe that this licensing board can and should grant a construction permit to the co-applicants."

Joe Farris had obviously expected this little duet. "We are not surprised at the similarity of the evidence that is going to be offered by the Staff and the Applicant." he responded. "Our point

Black Fox Hearings Suspended

The first phase of nuclear licensing safety hearings for Public Service Co.'s Black Fox Station ended Friday on a to-be-continued note.

The hearings which have lasted nine days, with two night sessions and a Saturday session for limited appearance statements, are expected to resume later this year.

"We are facing at least two more weeks of hearings," said Dow Davis, Nuclear Regulatory Commission staff attorney in the Black Fox case.

Davis said the NRC went into the hearings with about two ...sses, and only half of

TULSA TRIBUNE — OKLAHOMA, THURSDAY, OCTOBER 12, 1978 — 92 PAGES—8 SECTIONS

David versus Gol...

is that after the design has been cast in concrete, the changes … are going to be impossible to make."

He said we would prove that the board would not be able to find any solution to the "details" Dow had referred to and complained that the board had "unduly restricted the scope of this hearing."

Referring to Chairman Wolfe's invitation to the audience to "stay around for the evidentiary hearing and be informed," Joe maintained that if they were to be informed only on the board questions, the audience would not be "going to be informed at all," that they needed "to see the whole picture rather than a very narrow board question."

Knowing it was futile, he once more asked the board to reinstate each of the safety contentions.

Vaughn Conrad introduced PSO's Preliminary Safety Report. The NRC's Dr. Thomas reported that they needed one further piece of information, this one about breaks in a steam line that could cause loss of coolant (and a possible meltdown).

Wednesday morning's hearing began innocuously enough with testimony on PSO's financial standing. Our confidence that the board would judge them able to construct Black Fox was fully justified.

Then Joe Farris asked GE's Gerald Gordon to explain "for the benefit of the public here" the term, "intergranular stress corrosion cracking" in relation to plant safety.

During the ensuing discussion, Joseph Gallo complained bitterly, "Mr. Chairman, I thought Mr. Farris was representing a set of intervenors led by an organization described as CASE, and not the public. He seems to be continually playing to the public, whoever that is."

A voice from the audience spoke up, "We're the public. We're interested." Chairman Wolfe warned that they couldn't participate in the evidentiary session.

For years, the NRC and the nuclear industry had been using a statistical analysis called the Rasmussen Report[1] to support their contention that nuclear power was safe. GE's witness, Dr. Gordon, had told us that the report included an analysis of the probability of pipes breaking from intergranular stress corrosion. Joe then asked him if he were aware that reactors of the kind PSO had contracted for had seven to ten times as many such cracks as the Rasmussen Report assumed.

He wasn't. Nor did he know whether intergranular stress corrosion cracking was one of the 27 safety-related items identified in the GE Reed Report. Nor did Dr. Gordon know what else was discussed in the Reed Report.

I have heard it said that in a hearing a clever lawyer never asks a question to which he does not know the answer beforehand. He's up to something, I told myself.

Until 1978, I had been convinced that Tom Dalton was the most brilliant lawyer I had ever met, but when I saw Joe Farris in action I knew his ability rivaled Tom's. Joe simply wanted to get out on the table the point that nobody outside of GE—and few inside—actually knew what was in the report and that without a thorough knowledge of its findings, the NRC was working in the dark.

Joe's continued questioning about the Reed Report finally goaded the NRC's William Paton into objecting that "I don't believe it is in evidence before this Board. I don't believe that anybody has given this Board any competent evidence to connect it with anything, much less the present question that is before this Board. I think interrogation about the Reed Report is not appropriate."

Joseph Gallo chimed in: "The intervenors in this proceeding specifically attempted to introduce the Reed Report, and what it contained, as a contention in this proceeding and it was dismissed by the Licensing Board."

Now Joe Farris had them where he wanted them! "If intergranular stress corrosion cracking is a part of the GE Reed Report, I think that would bear very heavily on ... whether GE is committed to remedial measures," he suggested gently.

Chairman Wolfe, who had not attended the prehearing conferences, was looking puzzled. I held my breath.

"Mr. Gallo, you say the Reed Report and/or allusion to that report was rejected by the Board? At what time was that, Mr. Gallo?"

I could breathe again. He had swallowed the bait!

Chairman Wolfe asked for a copy of the June 29th ruling. I held my breath again.

It's a wonder I didn't turn blue in the hour that followed. A fierce argument ensued among the three sets of lawyers—couched in the politest language possible. Shortly thereafter, Chairman Wolfe told us that the board had already been discussing the issue, "more particularly since we reviewed Mr. Hubbard's proposed written testimony."

And here Mr. Wolfe made good on his commitment in the opening session: The board decided that they needed the Reed Report, partly because "it was brought to our attention again during the course of the limited appearance statements."

In this episode timing was, of course, everything. If the existence of the Reed Report had not been made public knowledge, if NRC commissioner Kennedy had not started—just that week—to ask publicly why it was not available at least to the NRC, if Mr. Hubbard had not been available to us, and if so many of us had not insisted publicly on the report's importance it would still be moldering, unread.

Because there was proprietary information in the document, Chairman Wolfe was careful to add a caveat: "When we proceed to take testimony on the Reed Report the proceedings will be *in camera*. That is understood."

"How long," he asked, "do you think it will be before the parties can informally get together and arrive at the wording of a protective agreement and … order for signature by the Board?"

Joe seemed to swell to twice his normal size with the effort of keeping his composure. Jubilantly, he pulled his carefully prepared rabbit out of his hat: "Mr. Chairman, I have a copy of the form agreement that was drawn up by the New York County of Suffolk, an intervenor in the Long Island Lighting Company hearing. It is a form agreement as to disclosure of confidential commercial information, in particular the Reed Report, which I don't believe has ever been executed. We could change that form agreement for our purposes here and have that, I would think, by tomorrow, if the parties agreed to it."

On the front page of that Thursday's *Tulsa Tribune*, the heavy black headline stretched from margin to margin: "Black Fox: David Vs. Goliath." The story by Joe Howell and Bob Bledsoe read:[2]

The battle over the Black Fox nuclear power plant Public Service Co. of Oklahoma plans to build began to take on the aspects of a complex modern David and Goliath story today.

Cast in the role of a potential "David" was "Aunt" Carrie Dickerson, a Claremore farmer's wife who started Citizens' Action for Safe Energy (CASE).

CASE won a major victory Wednesday when the Nuclear Regulatory Commission Atomic Safety and Licensing Board handling the Black Fox application decided to issue a protective order designed to make the General Electric's Reed report available on a confidential basis for use by lawyers in cross-examining witnesses during closed-door sessions.

The reporters told of my background, my several careers and my children. Then they continued with their story:

She is a friendly person with a guileless, almost childlike face. She is neither pushy nor overbearing; a disarming smile is her weapon—she is not a Carry Nation.

After the licensing board had agreed to let her lawyers see the Reed report, she was grinning from ear to ear as she asked a reporter, "Have you heard about our breakthrough?"

When the reporter asked what she could do with a report she could obtain only in secret and could not tell about, she replied with an impish grin and twinkle in her eyes: "We're not worried about that."

This modern "David" now has the slingshot in her hands and, as with the little shepherd boy of long ago, has no fear her stones will miss their mark.

I appreciated all the nice things they said about me, but I knew that my role had been only that of the catalyst who set the stage for the work of hundreds of other people—including Commissioner Kennedy and the hearing board.

When Joe Farris was asked what he would do with the Reed report, he replied, "We don't know. We don't understand why PSO doesn't want to see the report. It may be that the twenty-seven items have all been taken care of and are no longer a problem, and it may be that when PSO learns about them it will decide to cancel the contract with GE."

"A summary would not serve the purpose."

Most of the third day of the hearing was spent discussing the safety of spent fuel pools and prevention of hydrogen explosions within the reactor pipes.

In my earlier investigations I had learned that because there was—and is—no safe way to dispose of nuclear fuel rods when their usefulness is ended, they are stored in large pools on the plant sites to await the hypothetical day of a real solution. Their structure is crucial; if the water were

TULSA WORLD, TUESDAY, OCTOBER 31, 1978

Foes' Efforts to Get Se

make the Reed Report available for inspection by the board in camera (a closed proceeding).

The motion also requests that the licensing board permit oral arguments before deciding the Reed Report issue.

Following the latest GE move, parties to the Black Fox hearings have until Nov. 7 to file written briefs with the Atomic Safety and Licensing Board.

The CASE position is that material in the report could aid in the safety

GE boiling water reactors. The company has said 27 items in the report were safety-related.

A memorandum filed in support of the GE motion states that during evidentiary hearings in Tulsa, "The intervenors entered a motion for production of the Reed Report, a proprietary 1975 GE product improvement study, which was not a safety review, and according to confidential reviews by the NRC and Congressional Committee staff, did not consider matters related to safety which were not otherwise previously known to the NRC staff."

dence t
port a
safety-
entire
format
matio
interve
"In
poena
the int
The
the he
ment
the la
ackno
"wa
Reed

to leak or evaporate from the pool, a disaster of immense proportions would ensue.

Most of the questions were so technical that it was hard to keep my mind on what was going on in the courtroom. I sat up and took notice, however, when a PSO witness mentioned that they might want to store more rods in the pool than the plans allowed for, because the prospects of long-term storage were so doubtful.

They know that? I thought. *Doesn't that tell them something about what they're doing?*

Joe Howell's story in the Friday *Tribune* observed that time and experience on the side of nuclear power promised to overshadow the intervenors' pleas. Knowing Joe, I shouldn't have been startled to read his next words:

> The real issue is time. The opponents feel that if they can delay the plant long enough, PSO may give up on it or public sentiment may become so strong that it will force a halt to the building of nuclear plants.
>
> In the three days of the hearing this week, the licensing board already has fallen one day behind schedule and has no hope of finishing the health and safety hearing when the time it has allotted for it runs out next week.

We hadn't realized our objectives had been so obvious. We had no intention of allowing Black Fox to be built, but if it were built, we intended that it be built with every safety precaution possible. We would leave no stone unturned to make it the safest nuclear power plant ever built.

The second day of testimony, former GE engineer Dale Bridenbaugh had given PSO something new to think over in his testimony for CASE. "There appears to be not enough business for all four nuclear power plant suppliers, and GE may be the one to drop out of the reactor business. If that

t Reed Report

the Reed Re-
s to the '27
much less the
constitutes in-
d lead to infor-
to any of the
ons.
pe of the sub-
eds the scope of
tentions ...."
was described at
,000-page docu-
an for GE. But
ndum from GE
this information
accurate. The
consists of a 21-

main report of some 140 pages ...."
Contacted early in October, a GE
spokesman in San Jose, Calif., said
he doubted the NRC has to powers to
force GE disclosure of the Reed
Report.
Dr. Glenn Sherwood, manager of
safety and licensing for GE, said
that the company in May 1978 turned
over a package to the NRC which
describes the 27 safety items and
gives a summary and status report
on each one.
According to Sherwood, to make
the entire Reed Report available to
the public "would tip off our compet-
itors on GE plans and strategies af-
fecting reactor sales."

should happen, GE's top employees would find employment elsewhere and GE would not have as able a staff available to work [on] solutions to [design] problems."

In a way, that was good news: If fewer plants were built, we would have fewer consequences to suffer. In another way, it was terrible news: We knew that once a reactor was on-line they wouldn't shut it down just because there was no competent engineering support.

NRC witness Bernard Turovlin compounded my consternation when he asserted that it was PSO's responsibility, not GE's, to have the plant built to NRC standards and requirements. Later in the hearing, Executive Vice President Martin Fate echoed Turovlin's revelation: "PSO has the ultimate responsibility for the safety, reliability and availability of Black Fox Station." The fact that he did so in the context of promising to spend two million dollars more for heavier stainless steel pipes to minimize the danger from hydrogen explosions was cold comfort.

One reassuring piece of news came out in testimony about an excavation at the site: It had revealed that what we had feared was a fault at the site was in reality just a depression in a seam of coal atop a perfectly sound layer of sandstone.

On Monday, Joseph Gallo reported that GE would provide excerpts from the Reed Report dealing with the twenty-seven safety questions and provide the full thousand-page text to the licensing board so they could check it against the summary for accuracy. The complete text would not, however, be available to the intervenors' attorneys or witnesses.

Joe Farris immediately objected. The thousand pages were themselves a summary of the original five-foot shelf of reports, which GE had supposedly destroyed. Joe contended that in making a summary of a summary, GE would have to draw conclusions and leave things out. There would inevitably be "some glossing over, some mis-characterizations."

Chairman Wolfe agreed that "A summary would not serve the purpose of allowing the intervenors to cross-examine fully and intelligently ... " and signed a subpoena, effective October 30, for GE to deliver the Reed Report to CASE attorneys.

(Lest I get my hopes too high, Joe Farris warned me to expect that GE would move to quash the subpoena.)

Richard Hubbard's written testimony had influenced the board's decision to subpoena the Reed Report, and the day before the scheduled end of the hearing, we finally heard him cross-examined. Lawyers Paton (for the NRC) and Murphy (for PSO) did their level best to prevent him from giving any information for the record other than "name, rank, and serial number." They were so intent on preventing him from saying anything substantive that at one point Chairman Wolfe got fed up and addressed a little homily to them:

> Yesterday the board observed, and today the board observes ... too much persistent objecting to questions that in fact are not objectionable. We're trying to proceed here to get the facts; ... you're not proceeding before a jury. ... We [the board] can pick our ways through the minefield ... So let's calm down and proceed with some efficiency.

Hubbard had been manager of quality assurance for GE's Nuclear Energy Control and Instrumentation Department. He supervised a staff of 150 people and was thus in a position to know that GE was selling its customers a pig in a poke. If they saw the pig as it truly was, they would not buy it.

However, he made it clear that the shortcomings were not all GE's. He quoted a General Accounting Office study asserting that "Although the NRC is responsible for assuring that nuclear power plants are constructed safely, it has not been independently testing the quality of the construction work." It seemed to me that he—and the GAO—were saying that the NRC had, up till then, been little better than a rubber stamp for both the reactor manufacturers and the power companies.

He said in so many words that quality assurance based on the PSAR and SER was not quality assurance. "The board should order the applicant and the staff to amend the PSAR and SER to reflect analysis, review, and documentation," he admonished them.

The issue on which he was chiding GE and the NRC, cracks in pipes and nozzles transporting water to cool the reactor, was not a minor one. On the contrary, "Without the water in the cooling system, the reactor core would drastically overheat and possibly melt its way down through the containment vessel and cause disastrous dispersal of radioactivity."

The hearing which had lasted nine days, with two night sessions and a Saturday session for limited appearance statements, recessed on October 20. It had earlier been assumed that the hearing would be completed on October 20, but it was now clear that that assumption was rash. Because only half the two dozen witnesses had testified, Dow Davis lamented that they were "facing at least two more weeks of the hearing." There were still a number of major issues to be dealt with—but the Reed Report was what we were all waiting for.

It is always difficult to retain a reasonable perspective and avoid the blindness of fanaticism, while keeping the faith against severe odds. I am no exception to this rule, so I was grateful for every omen that even hinted at success. During the fall of 1978, such omens were not hard to find.

In an interview, Deputy Energy Secretary John O'Leary called nuclear power a "has-been," observing that "the slump in the nuclear industry has come about mainly because of high construction costs, economic recession and little demand. Unless this is reversed, nuclear power will not expand significantly for a long time—if ever."

In the December 6, 1978, *Tulsa World*, Bob Mycue described PSO's plight: They had expected a Black Fox construction permit in 1977, four years after the announcement of their plans, but it was now the end of 1978, and the hearing would certainly extend into 1979.

"Although PSO has forced itself to adopt a patient attitude," he said, "there have been signs that all had not been well with the nuclear industry. The chairman of the Atomic Industrial Forum (the industry's trade association) reported that only four nuclear power plants were ordered in 1977, and none in 1978. Furthermore, eight orders or options for nuclear power plants had been canceled in 1977, and another five in 1978. Compared to the forty-one orders placed in 1973 (more than in any other year), and twenty-six in 1974, this had caused great concern to manufacturers of nuclear reactors and components."

We will win! I told myself once more.

(Kay Drey, who fought the Calloway plants in Missouri for years, tells me that as of November, 1994, only three of the forty-one orders placed in 1973 were ever completed, and in fact they were the last to be completed in this country. One of those was the Wolf Creek plant.

Another was the Calloway plant (although because of the success of Kay and her co-workers in getting an initiative petition passed, only one of those reactors was completed). And the third was the three reactors of the Palo Verde plant—the largest in the nation. Since 1976, there have been no new orders for nuclear reactors in the United States.)

The Reed Report

Our limited appearance statements had borne such rich fruit during the first set of the health and safety hearing that when the hearing reopened six and a half weeks later,[1] PSO had clearly decided to borrow some of our tactics. Thus began what, in retrospect, was a very funny "Battle of the Gallery." Entering the Federal Building courtroom, one saw on the right seats filled with people wearing "Support Black Fox" badges, while on the left the gallery was packed with "Stop Black Fox" badge wearers.

Despite all my anxieties about fundraising, Robert had helped me keep my sense of humor—and there really were some very amusing things to be seen, if one had eyes to see them. That first day, we were treated to the hilarious sight of Joe Gallo piously quoting Chairman Wolfe. "This is not a trial," he admonished Joe Farris after one of the latter's objections. Unfortunately, he

Benjamin Spock & Carrie

Baby doctor Benjamin Spock and wife lead Tulsa prote

Dr. Spock urges Tulsc any method to block

By YVONNE REHG

Dr. Benjamin Spock, baby doctor turned anti-nuclear power activist, told Tulsans Tuesday night to stop construction of nuclear power plants possible — legal or

"Nuclear scientists now know that there is no level of radiation so low that it is not going to do harm," he said. "The government admits that we don't know anything about how to store it indefinitely."

Spock urged his supporters not to be "intimidated by the govern-

spoiled the effect when he went on to declare, "The credibility of witnesses is not important."

One interchange was especially silly. It involved Joe Farris, who was drawing things out shamelessly, Joe Gallo, who was incensed, and Dow Davis and the hearing board, who were rapidly lapsing into hilarity themselves:

> *Witness:* Each time you have a demand, you have a probability for failure on that demand.
>
> *Farris:* If I am reaching—
>
> *Witness:* (interrupting) If you increase the demands, you have twice as many demands times probability per demand.
>
> *Farris:* If I reach into a deck of cards for the ace of spades, the more I draw, the more chance there is that I will get it as the pack dwindles.
>
> *Witness:* Not per draw.
>
> *Gallo:* Objection! Unless Mr. Farris is prepared to connect up his

Tom Mooney with a PSO supporter

to use

plants

followers to try to get
"stopped before it gets

ahead of time if you can.
companies are more vul-
you can get them before
the damned things."

Dr. Spock & Mary Morgan lead a parade of demonstrators

card trick to the question of the probability for scram reliability, I think this is irrelevant.

Farris: I think the board can take judicial notice of the number of cards in a deck and the number of aces of spades in a deck of playing cards. I think that is common knowledge to the witness. The amount of time the board has been spending in Tulsa, I imagine they have had occasion to look at a deck of cards.

Davis: The staff agrees with the applicant that the foundation for the analogy hasn't been set.

Wolfe: The objection is sustained. To the extent the witness may have begun to answer the question, that testimony is stricken.

And no, we haven't played cards.

(general hilarity)

Shon: There are far too many other interesting things to do in Tulsa.

Farris: Wait until you've been here five years, Mr. Shon!

Other things were not so funny—beginning with the testimony about the emergency core cooling spray system, which was designed to cool the reactor in case the pipes carrying water ruptured and the temperature began to rise. It turned out that not only had such a system never been tested properly, the partial testing that had been done had indicated that it might not work![2]

In some ways, the testimony that went on outside the courtroom was just as important as that inside. Black Fox manager John West confidently told reporters in the lobby that "in the event of a malfunction or an accident, the plant operator will be able to take remedial measures to mitigate the consequences and protect the health and safety of the public."

Reporters also questioned former GE engineer Greg Minor about the fire at Browns Ferry. Greg had helped design the control room, so he knew what he was talking about when he told them that, "If safety systems had worked as designed, the reactor shutdown should have been accomplished in a few minutes." Instead, it took fifteen hours and prevented a meltdown with only minutes to spare.

The fire "came as a real shock" to Greg, because the plant design had incorporated the best engineering then available. "If the good plant gets into trouble," he asked, "what about all the others …?" Although he had been working for years with the same lethal technology, it was not until Browns Ferry that he began to question seriously what he was doing.

"It is my belief," he concluded, "that engineers cannot design to overcome human error. Humans are fallible. That includes engineers, plant operators and maintenance people."

The following day's testimony reassured us that—in a new plant—the same problems that had nearly caused a meltdown at Browns Ferry would not happen again. However, they did nothing to reassure us about problems that had not been addressed—or even recognized. Would it take a tragedy to identify them?

One of those that had been identified—and that could cause a meltdown—was an "anticipated transient without scram." That is just a fancy way of saying something—like, for instance, a power surge—that would cause an accident and at the same time make it impossible to shut down the reactor. (I suppose that such jargon saves time for engineers, but it sure makes it harder for ordinary folks like me to understand!)

GE's witness wanted to ignore the whole thing, because he felt that such a happening was extremely unlikely. The NRC's A.C. Thadani disagreed, saying that his conclusion was based on inadequate methods. "PSO should modify the design and construction of Black Fox in whatever way that's necessary to deal with the ATWS problem," he said unequivocally.

Greg Minor supported him. "… ATWS [is] one of the primary causes of accidents—primarily in BWR's [Boiling Water Reactors] …" and suggested several possible modifications.

The modifications would have been very expensive, so John Zink resisted: "… Public Service does not believe that ATWS is a concern."

Three days after the hearing resumed, I sent out a tabloid newsletter, *The Soft Path,* to CASE supporters. Its name came from Amory Lovins, the FOE expert on energy, who called renewable energy the "soft path" and oil, coal and nuclear energy the "hard path." Our tabloids had to be

printed at newspaper offices, but because some of the newspapers supported Black Fox, a number refused to accept our business. Often, we
had to go to Okmulgee or Pawhuska to get them printed. We would get
permission to use the paper's typesetting facilities, and sometimes it took
all night to get everything ready to print. Eddie Bryant, Nancy Perreault,
and John and Helen Hoogewind were among those members who
helped with this.

One time the equipment was out of commission at the *Okmulgee
Daily Times*, and we had to go elsewhere. We found a newspaper office
at Tahlequah that could do it, but when Eddie Bryant arrived with the
copy and they realized that we were anti-nuclear, they refused to do the
job. We telephoned all around and finally found a newspaper at
Pawhuska that would do our printing.

In spite of his stance against my work, Don Dodd at the *Claremore
Progress* agreed to allow us to have the printing done there during the last
couple of years of the fight. "Your greenbacks are as good as anyone's,"
he told me. He even let me use their typesetting equipment after Pat
Reeder had very patiently instructed me in its use. It was a great relief to
be able to work so close to home instead of traveling long miles to do the
preparation and printing.

In the newsletter, I thanked everyone for their appearances at the
hearing in October and for their part in getting the Reed Report subpoena. Once more I explained how much our expert witnesses had
meant in the hearing process and told of the wonderful work our attorneys were doing. Once more I said how important it was for us to raise
enough money to continue with the hearings and how I believed we
would eventually win—but that if we dropped out of the hearings, we
would have no chance to win. Then I invited everyone to a benefit lecture by Dr. Benjamin Spock on December 12 at the Church of the
Resurrection. (Its pastor, Father William Skeehan, was a faithful ally
throughout our struggle, and his poetry had enlivened the limited
appearance hearing.)

It was a great thrill for me to have this tall, gentle, distinguished-
looking gentleman—still handsome and with a charming smile and
manner—here in Tulsa. When I was a young mother, I had referred to

his *Baby and Child Care* during the many alarms and puzzles of parenthood—especially when I was trying to uphold some newfangled notion in the face of my mother and mother-in-law's certainty about the efficacy of an "outdated" remedy—and later I was to use it as a textbook in my high school home economics classroom. By the time I met him, he had become a kind of living national monument, and it was rather like welcoming George Washington in person.

It was also a thrill because his name had the power to draw attention to our cause from people—especially some reporters—who would never otherwise have given us the time of day. Because of him, people across the country became aware of what we were doing and why.

Accompanied by his beautiful wife, Mary Morgan, he told a crowd of about 200 CASE members and supporters, "I don't want the children being raised by my book to be given cancer and leukemia because some people believe nuclear power is the easiest way to make electricity."

CASE and Sunbelt members from across the state had already been organizing a demonstration for the morning of December 13, and we invited people attending the lecture to join us—and Dr. Spock—in our march through downtown Tulsa. We had prepared badges saying "Give us the Reed Report!" for everyone. Our posters shouted the same thing, adding, "Stop Black Fox," and "No Nukes." Policemen were stationed at all the intersections we were to traverse, ready to stop traffic for us.

About 200 people met in front of the Sixth Street entrance of the old East Central High School in Tulsa. PSO had bought it and turned it into a beautifully designed office building covering the whole city block. It was symbolically and geographically perfect for our purposes. Dr. Spock and Mary Morgan led our parade, and Ilene and I followed directly behind. Then came poster-brandishing pairs of Sunbelt and CASE members. We marched several blocks to the federal courthouse at Fourth and Denver.

When we reached that intersection, the three members of the hearing board were just approaching the crosswalk. They did not turn a hair or acknowledge by a blink of an eyelid that they saw us as they waited, with a growing crowd of other people on their way to work, for all 200 of us to cross the street.

We filed past a pickup truck waiting in front of the building, and each of us neatly deposited a poster and then walked sedately into the federal building. It was a cold, blustery day and we welcomed the warmth of the courtroom. It quickly overflowed to standing room only.

By the time we arrived, the "Support Black Fox" people had already taken their half of the 120 seats in the courtroom. We packed the other half—and the aisles and the corridors of the courtroom—so tightly that it was difficult to breathe. We were so excited, waiting to hear the outcome of the Reed Report subpoena, that we could hardly breathe, anyway. Dr. Spock sat beside me in the front row, listening intently as the drama unfolded before us.

In the final, three-hour session of the hearing on December 13, 1978, GE's special counsel, George Edgar, summed up the company's position on releasing the Reed Report. "Before we are accused of compromising safety, let me hasten to add that ... GE has offered the NRC licensing board the opportunity to examine the Reed Report in order to satisfy itself as to its contents. GE is not obligated to give the intervenors anything."

He went on to explain that one of the major reasons they didn't want us and our witnesses to see the report itself was that our witnesses were former GE employees. They were opposed to "nuclear power in general and GE's participation and effectiveness in the nuclear power plants in particular."

Dennis Dambly, an NRC staff counsel, supported the subpoena: "The Reed Report is relevant to some contentions in the Black Fox proceeding."

"Can you represent to the board that ... the Reed Report can have no possible bearing on health and safety issues [in the Black Fox case]?" Chairman Wolfe asked Mr. Edgar.

"I cannot represent that to you; I cannot," he replied.

Joe Farris, handsome and personable, clearly enjoyed his moments in the limelight. He had come into the hearing that day prepared to indulge his taste for hyperbole. And he did, declaring:

> We ask that you let [the Reed Report] be exposed to the harsh

light of cross-examination to see it for what it is. If it turns out to be an innocuous hobgoblin, then we can put the controversy behind us. But if it turns out, like Dracula, not to be able to stand the light of day, then we ought to put the wooden stake in its heart and deny the construction permit for Black Fox Station.

At this point, we couldn't contain our admiration. We applauded and cheered. "If there are any more outbursts," scolded Chairman Wolfe sternly, "we will clear the courtroom." Unwilling to miss anything, we quieted down.

More seriously, but no less dramatically, Joe asked the board to weigh the public's health and safety against

> ...the vague, ill-defined and unproven—and I emphasize unproven—concerns of General Electric that its competitive position will be harmed ... I don't find [the board suggestions] acceptable for one reason, and one reason only:
> The Board has at its disposal Dr. Purdom and Mr. Shon, who are engineers, scientists. The staff has at its disposal engineers and scientists, and a lot more experience than Intervenors' counsel do. The same goes for the Applicant and General Electric.
> If it is put upon me to decide whether or not a summary is an adequate characterization ... of the Reed Report itself, I am afraid I may do a disservice to my client.
> ...we retained Messrs. Hubbard, Minor, and Bridenbaugh to give us that technical expertise and to help us focus on these issues ... We need their input in order for me to do the best job I can for my client, and they need to see the report in order to give me that input so that I can cross examine whether or not General Electric's ... design is sufficient.

The board conferred for several minutes and Mr. Shon reminded Joe that, "the board's proposal ... did not involve your consultants looking ... at the Reed Report." Then he pointed out that the proposal offered, as safeguards, that Joe and our other attorneys could see the document itself, that the consultants could see the summary. Other prehearing conferences or hearings would be allowed if he "as a very competent and now technologically oriented attorney felt [it] needed further clarification."

Joe shrugged off the flattery gracefully and then turned it against the proposal:

... I appreciate ... the compliment that I am now technologically oriented. If I appeared as such, it is because I had either Greg Minor, Dick Hubbard, or Dale Bridenbaugh sitting at my right side to tell me what the witnesses' answers meant when they answered them.

You may have noticed a blank expression on my face from time to time. That reflected the state of my mind ... but with those gentlemen there, the blank expression would on occasion clear up, and I could proceed with another question and probe deeper.

I fear the same thing will happen when I see a summary of what now already appears to be a summary. ...

Chairman Wolfe reminded him that GE had given him to understand that the Reed Report was the actual report rather than a summary. "Now, whether you agree with that or not, that is what GE has represented. Is that correct, Mr. Farris?"

Joe agreed, saying that GE had told him that the Reed Report was all that was available "now that the so-called subtask reports of 713 pages or so [have] been destroyed."

Taken aback, Chairman Wolfe asked, "You have no reason to disbelieve GE when they say they have destroyed the 713 page backup material, do you?"

"No, sir," Joe shot back. "I believe the lessons of Watergate have been learned well here."

The room erupted into pandemonium as spectators laughed and applauded.

Pounding his gavel on the table, Chairman Wolfe remonstrated, "There will be no demonstration, no applause in the courtroom."

Mr. Edgar asked, "How," if Joe Farris could not accept the fact that GE could produce an acceptable summary, "can he accept going into an *in camera* proceeding and relying on the board for advice?"

Chairman Wolfe wanted to know whether or not Joe trusted the board to pick up on points that he might have missed.

Joe hastened to reassure him:

I don't think I ever have questioned the integrity of this board. Certainly some of the most probing questions that have been asked in this proceeding have come from this board.

Bridenbaugh, Hubbard, and Minor were working at GE when the Reed Report was prepared, and Mr. Hubbard, at least, had some direct input into a portion of that report. Whether that input is reflected accurately or not, only Mr. Hubbard can tell us.

I am sure the board would come up with some things that I didn't know, but I think you would have to admit it is quite possible that Minor, Hubbard, and Bridenbaugh would come up with some things that you didn't notice.

When the hearing recessed that afternoon, we were told it would resume in early 1979. We would have to wait until then for an answer to our plea.

PSO leaders knew that the Reed Report issue could delay a decision on the Black Fox construction permit for months. They had completed the work allowed under the Limited Work Authorization. Equipment and workers were idle. Interest on their loans was mounting. Inflation was increasing the cost of materials. They told the press that each day's delay was costing $50,000.

CASE members felt that PSO had been foolish to spend any money at all on the site, and that the NRC had been irresponsible in encouraging them to do so. Of course, every day of delay was to our advantage, but we had been entirely candid in telling them that we were going to stop—not delay—construction.

We knew that rather than the stockholders, the consumers of electricity would, in the end, foot the bill for the company's imprudence. We also knew it was better to pay to stop Black Fox than to pay for allowing it to be built. And I knew we would win: We had delayed Black Fox for two years already, and we were determined to delay it forever!

I didn't get my best Christmas gift that winter until after the New Year. On January 3, 1979, Joe Farris called me to say that he had reached a compromise with General Electric, and they would produce the Reed Report. We didn't get all we wanted. Our expert witnesses would be given verbatim extracts only on issues that had already been accepted as contentions, and the hearing would be closed to the public—which included the intervenors!—and the press. However, given GE's preference for keeping it totally secret, it was a tremendous victory!

Dow Davis called a Washington press conference to announce the news, and George Edgar (from GE) put the best face he could on their concession: "We felt it is important to avoid protracted litigation over this issue and to … [clear the way] for an expedient conclusion of the Black Fox hearings."

Joe Farris believed the information in the report would be the "single most important evidence ever adduced at a nuclear licensing proceeding. If the report has the same impact on the ASLB it had on me, the end result will be to deny the construction for Black Fox Station," but lamented that "… specifics cannot be revealed to those who worked so long and hard to see the report."

The Reed Report brought up a number of questions that Joe would have liked to enter in the hearing, but according to Chairman Wolfe's decision in December, we were not allowed to enter new questions. We were all intensely frustrated that neither had we known about the report at the beginning of the intervention process, nor could we take the best advantage of it when finally we received it.

What a relief when, later, Joe received a call from one of the members of the board telling him that in studying the Reed Report he had come up with twelve new questions and that the board planned to introduce them into the hearing when they reconvened on February 19, 1979.[3] The board members had come up with exactly the same questions that Joe had!

Frank Thomas at Kerr-McClellan Waterway, Catoosa

Crowd Braves R

Wolf Cr

"The banks won't finance them."

We learned that a Westinghouse reactor pressure vessel bound for Wolf Creek, due north of us, in Kansas would be journeying up the Verdigris River, preparing to dock near Catoosa at 8 a.m. on January 5. Seventy-five protesters braved the cold to rally against the shipment. Seventeen were arrested.

Nancy Perreault, both a CASE and a Sunbelt member, wrote,[1]

> We walk to the site of the protest along snow-covered railroad tracks, our boots crunching in the icy snow which is now packed and slick. Our breaths are little puffs of steam. Some of us are so bundled up against the cold that it is difficult to walk. Icy wind hits us—blowing in from the river. We bunch together as the newly risen sun shines through the clouds and climbs higher to shimmer on the icy river—gold, silver, green and white. We form a semi-circle around a tall Seneca ... holding the American flag. The medicine man asks a blessing. We stand waiting ... "Here it comes," someone says softly. Now we see the reactor vessel emerging,

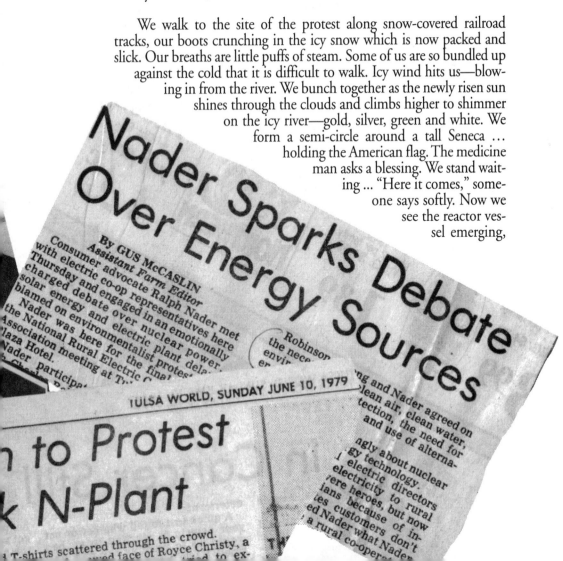

Nader Sparks Debate
Over Energy Sources

By GUS McCASLIN
Assistant Farm Editor

Consumer advocate Ralph Nader met with electric co-op representatives here Thursday and engaged in an emotionally charged debate over nuclear power, solar energy and electric plant power. Nader was here for the final protest ... blamed on environmentalist protes...

the National Rural Electric C... Association meeting at T... Plaza Hotel...

Nader participa...

TULSA WORLD, SUNDAY JUNE 10, 1979

...to Protest
...k N-Plant

...T-shirts scattered through the crowd.
...face of Royce Christy, a
...to ex...

moving slowly around the bend of the river.

"You're just a bunch of bums. Troublemakers! You just want people to freeze in the dark ..." shouts a man from the crowd. No one
pays the slightest attention to him.

The protesters stand silently waiting together until they are
placed under arrest, handcuffed, and led, single file, through the valley.

At the time of the protest, the Wolf Creek owners were meeting with
NRC officials to discuss construction problems at the plant, including possible substandard concrete in the reactor base. However, the 330-ton reactor vessel continued its journey on a special twenty-two-axle railroad car.

Kansas protesters tried once more to halt the vessel, as it approached
the site. Francis Blaufuss lay down on the railroad track as the train carrying the vessel approached. Francis and his brother Tony were arrested along
with their thirty-four companions.

Several months later, 1000 mostly-young people from the Sunflower
Alliance and Kansans for Sensible Energy, stood in the rain, some under
umbrellas or makeshift shelters, to listen, clap and sing, "We shall not be
nuked." Tulsa CASE member Harry Allison was there, representing the
Union of Concerned Scientists. Harry wanted to go on record as opposing
nuclear power. Courageous and heartfelt as these demonstrations were,
they were insufficient in themselves, and the legal intervention was under-
funded. Wolf Creek went on-line in 1985.

Going up against "the system" was never easy. When Ralph Nader
came to speak at the Lloyd Noble Center at OU on citizen action groups
and their impact on issues ranging from tax reform to nuclear energy,[2] I
looked forward to hearing him again. Armed with anti-nuclear brochures,
I established myself beside the sidewalk and handed a leaflet to everyone
who passed on the way to the entrance. Soon a guard came along, told me
I was not welcome, and ordered me off the premises. I pretended to comply, but as he entered the building I settled myself on the opposite side of
a large evergreen tree and continued to hand out leaflets. Presently a CASE
member came out looking for me. Ralph was asking where I was and
telling people to find some of my leaflets and put them on the table beside
his.

By the time I sat down in the front row and opened my notebook to take notes, he had already begun his lecture. When Ralph asked for questions at the close of his lecture, I held the notebook up and asked for his autograph. I was puzzled to be answered by a collective indrawn breath followed by a dead silence. In a bemused voice, Ralph muttered "Sure," and took out his pen. That autograph is one of my most prized possessions—the more so because I learned later that before I had come into the building he had said in response to a question that he charged a thousand dollars for his autograph! He never did send me a statement.

Unfortunately, I was unable to attend when Ralph came to Tulsa, on January 12, to urge participants in the National Rural Electric Cooperative Association's annual directors' conference to abandon nuclear reactors as an energy source. However, friends who went were invited to his room afterward. The pep talk he gave them was just the right boost to their spirits.

Reminding his NRECA audience that of 200 plants proposed, 130 had been canceled by the companies involved, he declared, "Nuclear power is over. Why? Because the banks won't finance them—the Bank of America already has refused—and the consumers can't stand the charges." He pushed for the same tax, price, and research and development incentives for solar and wind power that the government was offering for other sources of energy.

At about the same time the NRC, while reiterating its claim that "risk of a nuclear accident is comparatively very small," repudiated portions of the Rasmussen Report. Three years earlier, its staff had welcomed the Rasmussen Report warmly, citing it repeatedly in testimony and publicity. They finally came to the conclusion that it was a fatal flaw for Rasmussen to ignore the critical assessments of responsible scientists when they concluded that the chances of a person being killed by a nuclear core meltdown were about one in five billion (equal to being hit by a falling meteor)!

This, of course, alarmed members of the U.S. House Interior Subcommittee on Energy and the Environment, which held a hearing in late January to discuss the implications of new evidence on nuclear power plant safety. At the request of its chairman, the late Morris Udall (D,

Arizona), University of California Professor H.W. Lewis headed a panel to study the Rasmussen Report. The study found that, although the method used to develop risk projections was valid, the report had been unjustifiably precise about them. Claiming that there were mathematical errors and other problems with the report, Lewis nevertheless said that it was the potential of a large and unfamiliar event that so scared people.

The Union of Concerned Scientists told the Subcommittee, however, that "official claims about nuclear power plant safety have to be regarded at best as unproven speculation."

Legal Costs Endangerin[g]

By BOB MYCUE
Of the World Staff

The legal machine is breaking own for Citizens' Action for Safe nergy, the organization spearhead- g the fight to halt construction of

this phase of the hearings since last September has been the Tulsa firm of Green, Feldman, Hall and Woodard.

The firm has written the NRC giving notice of withdrawal of services

Tulsa att Jr. repres vironmenta and has co in areas oth the procee

Legal counsel withdraws

Black Fox foes out

BY PAT REEDER
Progress News Editor

"We will just have to go ahead without legal counsel and will not be as effective," Carrie Dickerson, Claremore, leader of Citizens Action for Safe Energy (CASE), said Saturday after attorneys withdrew from representing the anti-nuclear organization.

She said the group will continue their fight against the Black Fox nuclear power plant without legal counsel if necessary after announcing legal counsel ___resent CASE in the upcoming

Rogers
"In our
we have
burden
several
citizens
past th
curren
Not
for wo
Daltor
pert w
Wh
less

"Will forty acres be enough?"

CASE had assumed a crushing financial burden in our quest to stop Black Fox, one that was especially heavy during the fall of 1978. We were having to come up with the money for Joe Farris and his colleagues who were working on the Health and Safety Hearing, and Tom Dalton was still working on several appeals before the district courts. He was appealing the limited work authorization, the pollution discharge permit, and the water allocation permit. This meant that for several months CASE was paying a double set of attorney and witness fees.

Despite the confusion over the Sunbelt Alliance, CASE members contributed over $27,000 in the last three months of 1978. We divided the money among our attorneys and expert witnesses, but we still owed all of them money. Robert and I paid as much as we could of what was still owed, but eventually we had nothing left.

Tom was very busy appealing the permits that had been granted, but one time when he needed to prepare a brief for an appeal, he seemed to be dragging his feet. Because Tom had always been just as dedicated to the cause as we were, I couldn't understand what was happening, and I talked with a board member about the problem. "Maybe he's having trouble making ends meet," he suggested. "We're $2000 behind on his account, and he's probably having to do other work to pay the bills." I realized that he was probably right.

J-Plant Fight

ew T. Dalton
E in the en-
or Black Fox
ork for CASE
trial phase of

vention ever made in a NRC proceeding. We feel it is really unfair that Public Service Co. uses rate-payers' money to carry on its side of the Black Fox fight, the NRC uses taxpayers' money, and CASE

money

Inola.
p the construction of Black Fox, crushing financial burden. This specially heavy during the last even though the many concerned ve contributed over $27,000 in the we have not been able to stay penses," she said.
SE owe the legal firm about $19,000 ety hearings, but they also owe Tom entation in the water issues and exeral thousand dollars more, she said. kerson said she feels CASE will be roun will continue their fight when umes safety

Who, I wondered, could save the day? I racked my brain—and my address book, and the only name I could come up with was a long-time friend and fellow activist whom I shall call Benita. I had met her years before when I was operating my bakery and she was a member of the Organic Gardening Club and a customer of mine. Benita, who was very socially concerned, had contributed smaller amounts over the years to CASE. Might she be able to help us with this substantially larger need?

I phoned a mutual friend, Betty, who lived near her and asked her what she thought about approaching Benita. "Oh, it would be no hardship for her," Betty assured me. "She can well afford to help."

I phoned Benita and explained the problem and its urgency. "Can you do anything for Tom?" I asked.

"Why bother?" was her exasperated reply. "You're not going to win anyway."

Her unresponsiveness made me feel like a leper begging at the gate, but I knew that I had no choice but to try to change her mind. Like so many others, Benita was burnt out with the length of the struggle and the overwhelming odds against our success. If I were to succeed, I would have to find a way to help overcome her own despondency.

I spent over an hour on the telephone, pleading with her, trying to infect her with my hard-won and harder maintained faith. "We may win if we keep fighting, but we can't if we don't even try. If I didn't believe we could win, I would not have started the fight," I reminded her. "Robert and I would not have spent all our money and energy and six years of our lives."

Little by little, Benita's voice thawed and warmed up as she grudgingly allowed herself to be convinced that there was, indeed, hope. And finally she capitulated: "When do you need the money?"

Hardly daring to breathe in my anxiety, I told her we needed it the next day. There were six inches of snow on the ground and it was bitter cold, but now that she had changed her mind, her response was wholehearted. "I'll put on my boots and take it to the post office. It's only two blocks," she assured me.

Completely wrung out by the experience, I mustered enough strength to thank her properly and dragged myself limply to bed. The only comfort to be had there was the warmth of another body—not the sympathy I craved. But, in Robert's arms, I drifted into a sound and restful sleep. Robert was working so hard to keep us going financially that I could not bring myself to add to his burdens by sharing mine with him. He needed all his extra energy to keep himself on an even keel.

By the time we received the check the next day, I had recovered my aplomb, however, and I called Benita to tell her again how much her help meant to all of us.

Tom immediately began to prepare the appeal.

Early in January, Joe Farris' firm sent us a bill for $19,500. Because the balance was so large, Joe wrote to tell me that I would have to make a payment of $11,500 by January 21 for him and his colleagues to continue representing CASE. Joe apologized for the ominous tone of the letter, but said that he wanted to be sure I understood their position.

My initial response was panic. I had spent all our personal money, other women and I had made quilts to raise money, and I had humbled myself to beg for contributions, but no matter what I did, there was never enough. "God," I prayed, "tell me what to do now!" But no inspiration lightened my despair.

If you don't have inspiration, you have to make do with perspiration, so I pulled myself together and wrote letters soliciting contributions and held meetings where I explained that a bare financial cupboard could not feed the fight. It was January, and people had not yet had time to recover from the expenses of Christmas. Heating and electric bills were high, and people felt unable to dig any deeper. My efforts were unsuccessful.

The Claremore Chamber of Commerce chose this time for its annual gridiron roast. I was roasted—by furniture store owner Mickey Walker, who wore a dress and a large placard that read, "AUNT CARRIE"—and it couldn't have come at a better time. Mickey chanted, "I promise you all I will fast, but don't know how long I can last. If Black Fox goes on, I can't say just when, but some day, there'll be a BIG BLAST!" I laughed as hard as anyone else when I saw the picture of Mickey and the report in the newspaper. It was better to be teased than to be in Coventry—to be ostracized as I had been by so many Claremore people when I first started the fight. If anxiety added a hysterical note to my laughter, no one noticed.

January 21 came and went, and Joe had not received his $11,500. He wrote another letter withdrawing as attorney for CASE, effective February 3, 1979. I understood. He had been very lenient with us, but enough was enough! He needed his money. Understanding did not save me from writhing with mortification, however, when notices appeared in the newspapers on February 3 and the NRC board was told he was withdrawing as our counsel.

Some of the hurt was soothed when several law students and young attorneys telephoned to offer their services for the hearing that was to resume on February 19. How I wished that I could take advantage of their generosity! However, I knew that it had taken Joe weeks to prepare properly—and he was an experienced, gifted lawyer—so with gratitude and regret I thanked each of them and told them it wouldn't work.

Like Tom Dalton, Joe Farris had done us proud. It was exhilarating to watch his performances in the hearing room. He was always on his toes and ready with a quick comeback, fencing with wit and judiciously applied facts. He would set up courtroom scenarios as if he were scripting a movie. And he made sense! He studied and prepared thoroughly for each day's hearing, and we knew he was truly committed to our side. No one could take his place.

Kathy Groshong, as always, was a pillar of support. She asked nineteen people to guarantee $1000 each to the attorneys if we were unable to raise the money in six months' time. Carrying the list of guarantors, I went to Joe's office and told him firmly that we rejected his resignation. "I'm sorry, Carrie," he told me, "we can't go along with Kathy's idea. You may as well withdraw CASE now, because it's impossible to stop Black Fox."

"Never!" I declared, feeling more alone than ever. "If we quit now we'll lose any chance to win. What will it take for you to continue to represent us?"

"Collateral," was the answer.

When Robert decided to lend his support in the intervention against Black Fox, we had discussed the possibility that it might cost us the farm. It is dear to our whole family, not only because we have known every acre of it, intimately, most of our lives, but because it represents our hopes, sacrifices and triumphs. To Robert, especially, it was always the place where he was a free man, untrammeled by the demands and frustrations of society, subject only to the dispassionate caprices of nature.

Bought by Robert's father with a forty-year mortgage in 1910, the farm had been saved from foreclosure during the dust bowl and depression only by the bank's agreement to forego principal payments so long as the interest was paid. The land had always kept the family fed and sheltered, but when our two eldest daughters were babies, Robert's mother took care of them while I began teaching school as a means of obtaining cash to resume payments on the principal. We all had investments—emotional and financial—in the land.

Robert, however, was as incapable of being a dog in the manger as he was of being halfhearted about anything he had made up his mind about. "Carrie," he told me, "if the farm is contaminated with radiation, it will be of no use to us or anyone else, but if we win, it will still be a productive place for someone to live on and farm."

That had been before Robert's surgery, though, and he was not well when I went to see Joe. I could not bear to add to Robert's sufferings by suggesting that we mortgage part of the farm.

In my lowest ebb, I prayed, "Please, God, tell me what to do." And suddenly I knew the answer: Uncle Sinnette's farm. "Will forty acres be enough?" I asked. They agreed!

I could have shouted with joy! One more insurmountable obstacle had been defeated. We would win! Summoning all the dignity I could muster, I thanked Joe and his colleagues and promised to return on Monday morning with the abstract and land description. As I left, I exhorted them to "be prepared to do the best job you have ever done—and keep a positive mental attitude. We will need it to win!"

When I returned, Joe and three of his colleagues and their assistants had worked long past midnight throughout the week-end to prepare for the hearing. Those extra hours increased our debt, so I mortgaged forty acres of Uncle Sinnette's farm for a total of $26,500.

The land that saved that particular day is a story in itself. For years after my father died, I had little contact with his people because they lived so far away. When I went to Oklahoma A&M College, however. I met a Barefoot cousin who was also a student there, and on occasional weekends I went home with her to Newcastle to visit my grandparents, uncles, aunts, and cousins. They were a closely knit, warm, and loving family, and they welcomed me like a little lost lamb.

One of the members of the family was my father's younger brother, Sinnette. Never married, he lived with his parents as long as they were alive and then moved in with my Aunt Vivian and her family. Sixteen years old when my father died at thirty-six, Uncle Sinnette had hero-worshipped his older brother; his death had been a terrible blow.

Uncle Sinnette looked a lot like my father's pictures, and I very quickly felt a strong kinship with him. After that, that bond was always there as one of the constants in my life. It gave me a sense of security that I needed, and as I continued to visit through the years, it was deepened by the long conversations he and I would have.

During one of those visits Uncle Sinnette told me he was leaving half his 160-acre farm to my twin sister and me. The other half was for Aunt Vivian's son, Paul Albert Bell, who had survived more than three years of atrocious treatment and starvation during his imprisonment in Japan during World War II. I had not even known Uncle Sinnette owned a farm, but I thanked him sincerely and told him I hoped he would live to be a hundred, because his life and his friendship were worth more to me than any amount of land he could leave to me.

Since he didn't live to be a hundred, however, I accepted with a grateful heart the loving generosity that had provided the collateral Joe Farris needed. I mortgaged my half of the undivided eighty acres. However, I felt a terrible sense of remorse for not telling my twin sister about the mortgage. I felt that I had betrayed her. But I did not plan to lose the land; I intended to work hard to raise the funds to pay off the mortgage.

My husband was never to know what I had done. No one but Joe and his colleagues knew until after Black Fox was canceled, and by then, sadly, Robert was gone. Clara generously told me there was nothing to forgive. The abstract of the land now contains one not-too-small piece of history of Black Fox—in the mortgage cancellation.

Lawyers' fees were only one of the problems that kept me on an emotional roller coaster. There were much more mundane aggravations. Because of the hearing boards' deadlines, we often lived on the brink of disaster, barely getting important tasks completed on time. For one appeal, Tom Dalton had taken longer than he expected writing the brief. It was due in Washington the next day, and when his part-time secretary left for the day, Tom was still working on it.

Mary and I took our typewriter to Tom's office, and Mary used his typewriter and I ours while Tom finished the research and dictated page after page. It took the whole evening, and we still weren't finished: We had to make fifty copies and get them to Washington the next morning. In those days, there were no all-night duplicating businesses, but one of the CASE members, Pat Laner, owned a duplicating machine. Half-way through the evening, I phoned Pat and asked her if she would leave her door unlocked so we could get in when we finished. "I know it will be long after midnight," I apologized.

It was 3:00 A.M. by the time we arrived at Pat's house north of Claremore. When she heard us come in, she came to help copy and collate and stuff copies into the large envelopes Tom's secretary had already addressed. Mary and I drove back to Tulsa. We arrived at the office of the messenger service with minutes to spare, and their courier raced to the airport and ran through the door of the plane just as they were about to close it. Whew! It sometimes felt like being in a chase movie or an old-fashioned edge-of-the-seat melodrama.

It seemed as if we made millions of copies to send to Washington or Oklahoma City or somewhere else entirely. We usually paid ten cents a copy—although the actual cost of making them was less than three cents a page. Fees for lawyers and expert witnesses added up to a tremendous amount, but copying costs were also a big burden.

A CASE member finally persuaded her husband to allow me to use his office copy machine. Grateful to be able to cut our expenses, I tried very hard to avoid interfering with the work of his office. During the Environmental and Site Suitability Hearing, however, the job was enormous, and I spent so many hours there that it must have interfered with the normal work of the office. One day I worked from seven in the morning until seven in the evening, and I am sure some of the people in the office put off making their own copies for fear of interrupting me.

Another evening when I was there copying long past closing time, my friend's husband came in and found me still working. His face looked thunderous, and my heart fell as he came up to me. "I thought you would have finished this case long since," he barked.

Hoping to turn his thoughts into a calmer channel, I explained to him that hearings take a lot of time, especially ones that deal with complicated issues, and quickly went on to tell him how much I had appreciated his kindness through all the preceding months. He hardly allowed me to finish before he began a merciless tongue-lashing that left me in tears.

I knew I had been presuming on his kindness, but I couldn't understand why he could not have simply said something unequivocal but innocuous—like, "This has gone on too long. You'll have to find another place to make copies." Perhaps he had come in to deal with something that was worrying him into a panic and I was the unwitting spark that set off his anger. Perhaps he simply felt guilty about wanting me to stop and turned his guilt on me in self-defense. Whatever the reason, I was devastated.

After that evening, my only recourse for late-night copying was the TU law library. Commercial copy places all closed at five o'clock then, but the law library was open until midnight. At the end of the day, I often picked up the briefs that Tom Dalton had written and typed during office hours and then rushed to copy and collate them in time to get them to the Tulsa Central Post Office to be postmarked before midnight.

So much of this epic happened when we had snow. One night when I had barely completed copying at the law library in time to get the briefs postmarked before midnight, I started home over clear roads but with six inches of snow on the ground. As I was coming into Claremore, the car in front of me started weaving back and forth across the road, leaving me with my heart in my mouth. But then it straightened out and drove on. "He probably stopped in at the beer parlor back down the road before he set out," I told myself smugly.

Suddenly my car was making the same drunken gyrations. Only my car, instead of straightening out, took a sudden U-turn and landed in the snow with my door blocked by a utility pole. The snow must have melted during the day to form a layer of water on the road and then frozen when the temperature fell after sunset, forming an invisible sheet of black ice.

As soon as I realized that I was still alive and got my heartbeat back to something approaching normal, I pulled myself out of the passenger door and started walking toward town. When a police car approached, it might as well have been a knight in shining armor—I was saved!

The hero turned out to be a young man who had been in school at Claremore High when I had taught there years before. "Mrs. Dickerson," he demanded—his voice saying *Now I've seen it all* — "what are you doing walking alongside the road at two o'clock in the morning?"

"Before I explain," I begged him, "please radio in and have someone call my husband and tell him I'm OK. I'm sure he's worried sick wondering where I am."

When I arrived home safe and sound, Robert had had it. "This is for the birds!" he exclaimed in disgust. "There will be no more midnight rides in the snow." The next morning he took me to the bank, and we borrowed $6,000 to buy a cheap copier we could use at home.

"What does it matter if you lose your farm?"

On February 10, 1979, nine days before the health and safety hearing was to resume, PSO launched another media attack on the intervenors. Had all gone as they expected, their construction permit would have been issued in 1977. They told the *Tulsa World* it could not now be expected before mid-summer of 1979—a two-year delay.

I refused to accept any blame—either for myself or for CASE—for their financial woes or for risking future blackouts. "Just as we want power for the future," I said in my letters to the editors, "we want to make certain that we *have* a future!"

I continued, "Under intense questioning at one hearing, PSO had grudgingly admitted that they were planning to sell forty percent of the electricity from two coal-fired plants (then under construc-

Above: Helen Geary, Joyce Nipper and Marjorie Spees

TUL

General Asks Bl

...ing the plant, said they ...opies of the

half of the public ne lengthy licening h Service is operating tional construction ...ment $100 million ...t near Inola

10 SECTION C

Black Fox Behir

...de to the NRC's Atomic S and Licensing Board. Spokesmen for both the NR Public Service Co. of Okla

PSC Says Costs

"Public Service Co. has given official notification Black Fox Station nuclear project will be later than presently scheduled and th... escalating. The word came in hearings st...

tion at Oologah) to West Texas Utilities. They routinely sold up to fifty percent of the power they were then generating, to other utilities in cheap, 'sweetheart' deals. How," I asked, "could they justify these actions if the power was really needed so badly?"

"Far from doing a disservice to the people of Oklahoma," I declared, "CASE's intervention has meant that for the first time in a nuclear licensing proceeding in this country, safety information contained in General Electric Company's confidential Reed Report will be used in the hearing." I closed by reiterating that, "Our intervention will not delay but deny the Black Fox permit."

In some ways those seven days of the hearing in February were the most difficult I ever endured. At the other hearings, we may have had no real power but that of endurance, but we didn't feel as helpless. At the other hearings, we may not have known about hidden agendas or their consequences, but at least we could hear what people said and read what they wrote. At this hearing, there would be testimony about something especially worrisome or marginally hopeful, and the next minute the hearing would go *in camera*. I trusted Dale Bridenbaugh and Joe Farris to do everything that was humanly possible—but I wanted to see and hear them doing it!

We never knew how long those *in camera* intervals were going to last, so I had to gamble on whether I had time to do some essential grocery shopping or run an errand that had been put off far too long. When I returned, often as not the board and lawyers and expert witnesses would still be testifying behind closed doors, and I would have to gamble again on whether I had time for another errand or whether it would be better to try and find a telephone and make some calls.

As the days wore on, more and more often I found myself drifting with a changing cast of people to the City Cafeteria across

SATURDAY, APRIL 21, 1979

Fox Delay

nuclear station feels a need for the Black Fox licensing board to delay its decision.

"His apparent concern is from both a safety and financial point of view. The concern with safety apparently is that a similar accident could occur at Black Fox Station.

"From a technical point of view, we've explained several times previously that Black Fox is a fundamentally different type of facility ... Mile Island plant and ... any real im... ...tigation,'

TULSA WORLD, FRIDAY

chedule;

the street. Cathy Coulson-Currin, Ilene Younghein, Helen Geary, Marjorie Spees, Joyce Nipper, and Carol Oxley most often formed the core, and we would sometimes sit making plans in a vacuum or simply letting off steam.

In a strange way, those were almost the best hours of that whole nine years. We had an opportunity to get to know each other in ways we had not had time to before and to form a basis for enduring friendships that can be picked up and enjoyed after intervals of months or years.

That hearing also marked what I considered a betrayal by two of the Sunbelt Alliance members.

Undaunted by earlier refusals, Joe Farris began with a renewed plea to the board to open the Reed Report portions of the hearing to the public. To no one's surprise, his motion was denied. How grateful I was that we had Dale Bridenbaugh with the group, *in camera*, to see and hear and speak for us!

Dale Bridenbaugh was our star witness. When he was not testifying, he sat quietly at the table with Joe Farris, feeding him suggestions for questioning other witnesses. Dale and his colleagues had been testifying at hearings all over this and other countries. The differences between other hearings—full of passionate acrimony and contemptuous discourtesy—and the Black Fox hearings with their objectivity and civility, he found so striking that he made a point of it to the hearing board.

Indeed, he told them that that "lack of a neutral or objective attitude among his counterparts" was what finally prompted him to quit GE after twenty-two years. Dale also told the board that, to be fair, he had found that much as he distrusted American regulation of the nuclear industry, the situation in other countries was far worse. He said their plants were "carelessly" built and ill-run and -maintained by inexperienced workers dealing with shortages of replacement parts.

Dale felt our contentions were not as broad-based as they should have been—and as they would have been had he been involved at the outset. "It is unreasonable," he claimed, "to put the burden ... on an intervenor with very limited resources." He suggested that a more rational approach would be to review and, if appropriate, revise contentions periodically.

Our NRC attorney friend, Dow Davis, mourned that, "the technicality of the current health and safety hearing in Tulsa has driven even the most curious of interested citizens, into nearby coffee shops for relief. I really wish everybody could understand it all ... I believe having opponents in the hearing process has improved the design and safety of the plant and made the participants more conscious of their duties." (This from a man who, four years earlier, had tried to persuade us not to intervene!)

At the end of the first day's session, the board told us there would be a short period of testimony the next day, and then we would be told to leave so the Reed Report testimony could begin. It was then that, for me, things began to fall apart.

The next morning, I was surprised at the presence in the courtroom of two Sunbelt Alliance members who had hitherto shown little interest in the proceedings; it was the first appearance for one and only the second for the other. When Chairman Wolfe asked the public to clear the courtroom so that testimony on the Reed Report could begin, these two men rushed—amid the clanking of chains—to the railings that separated the lawyer' tables and the bench from the public area. Swiftly pulling those chains from their trouser legs, they padlocked them around the railing in a carefully planned act of civil disobedience.

The civility and respect that CASE members had so carefully maintained was—in one calculated blow—destroyed. In the minds of the people who had supported us because they trusted us to uphold, rather than defy the law, we had become lawbreakers—even though we were completely uninvolved in the action!

U.S. Marshals with bolt cutters freed the two Sunbelt members from their chains and arrested them—not for the first time—for disorderly conduct. If I had not been so despairing and angry at the disaster their behavior had perpetrated, I would have cheered at their words when the Sunbelt leaders made a passionate declaration upholding the public's right to know even trade secrets and commercial and financial information in cases of overriding public interest. They announced that they had "requested assistance from the governor's office, the attorney general, and the Oklahoma Corporation Commission" and filed a temporary restraining order to prevent closed hearings.

The hearing, two days later, on the temporary restraining order (which was denied) was attended by board members, attorneys and witnesses. In the meantime, a media pageant took over. The NRC hearing, as Dow Davis had complained, was dull and boring to the uninitiated, and the colorful respite of civil disobedience in the courtroom naturally attracted television, radio, and newspaper reporters in droves.

We had all been working desperately to raise money, because the $26,500 debt I had guaranteed with my forty acres had more than doubled during the hearing. CASE board member, Carol Oxley, had persuaded two of her friends to contribute $1000 each. When these people heard the media version of the "chaining," they vowed to have nothing more to do with our cause and rescinded their promised contributions. No amount of explaining that the acts of civil disobedience were committed by two Sunbelt members and had nothing to do with CASE made any impression on them. The Sunbelt members could satisfy the law with a few days in jail. We have spent years paying for their few moments of glory.

When they were released from jail, I phoned one of the Sunbelt members to tell him just what he had done. "I told you that while I would not join you in your acts of civil disobedience when you trespassed on the Black Fox site, I would do nothing to interfere, because I hoped with everything that was in me that you would succeed," I reminded him. "Could you not have had the courtesy to respond in kind?"

I explained to him that their court disturbance had cost CASE an immediate $2000 in contributions. And it certainly cost us thousands more. "If we can't pay the attorneys," I concluded in an attempt to awaken him to the personal implications of his act over and above those to the cause we both believed in, "I may lose my farm."

His reply shocked and disillusioned me. "What does it matter if you lose your farm?" he demanded. "*I* don't own a farm."

As I had suspected, the lost $2000 was just the tip of the iceberg. The woman who had earlier promised to be responsible for raising $20,000, came up to me during a recess to tell me how disgusted she was. "I'll not put any more money down this rat hole," she declared in a defiant tone of voice. "Anyway, you're not going to win," she defended herself from what she knew I was thinking, "so why bother?"

She was right. In sad bafflement, I was thinking, *The same old refrain! Where is your vision—your faith—your perseverance?*

Dale Bridenbaugh told us that when he had been a salesman for GE, long before the hearings began, he had been sent to Tulsa to make a presentation to PSO. There he spent the night on the "thirteenth floor of the Holiday Inn in beautiful downtown Tulsa while a tornado went through." That experience had alerted him to the need for power plant protection against tornadoes.

NRC's Kazimieras M. Campe agreed that "cooling tower tornado missile protection was inadequate," so PSO has pledged to "... provide tornado missile protection [by means of] a steel mesh grating over the cooling tower opening." Somehow, a fancy window screen didn't make me feel any safer from the horrendous power of a tornado.

Months earlier, when I first brought up the Browns Ferry fire in a prehearing conference, Vaughn Conrad told me that Black Fox's design included adequate protection from all kinds of fires, not just those from overheated electrical cables.

"Prove it to me," I demanded.

Dow Davis, who was at the same conference, went back to Washington and did his homework. Shortly thereafter, he sent me a copy of a letter to PSO confirming the validity of my concern and requesting a design change, and during that hearing Vaughn told us about the modified design. Silently, I was saying to myself, *Thank goodness I brought it up!*

I listened as patiently as possible to two more days of the usual round after round of interrogations, with testimonies that meant something—just not to me. But that wasn't unusual. I was so glad it all meant something to our expert witnesses and attorney. They understood!

Even that was sometimes cold comfort. GE, of course, wanted to be sure that no one who had inside knowledge of technical discussions or attitudes within the company would be given an opportunity to tell anyone what they knew for the record. And Dale Bridenbaugh, Dick Hubbard, and Greg Minor all had that knowledge. Therefore, even though Joe was prepared—and eager—to place Dale Bridenbaugh on the stand for direct testimony, PSO and the NRC had agreed that they would not cross-examine him, so he could not appear as our witness. In other words, it seemed to me, they were saying, "Not only do we not want to have to hear it, we don't want you to hear it, either." Once more, it seemed to me, the NRC was taking sides against the public!

Pick up any college engineering textbook and try to read two or three pages of the jargon-filled language you will find there. Only then will you have the merest inkling of what we subjected ourselves to in this hearing. The transcripts for those two days, including the testimony that the expert witnesses had filed ahead of time, were each more than three inches thick—enough to boggle the mind. Most of what was said dealt with small details and rarely mentioned their larger implications. It was as if I were to tell you, "Chairs set too near a stove may ignite," when you had no idea of the meaning of "stove" or "chair" or "ignite" and thus had no idea that you were being told that you could burn down the house that way!

So much of the testimony was irrelevant to my purposes, since it assumed that the risks of nuclear power could be reasonably borne. It was often difficult to keep myself from screaming, *You idiots! Why are you solemnly debating nonsense? Better yet, why don't you try subsidizing research in safe energy sources to the same tune as the Manhattan Project and Atoms for Peace? Then everybody could see how irrelevant all this is!*

However, I knew that while that kind of outburst might be momentarily satisfying, it would be the death knell to our hopes. All the people who had been saying I was just a crazy little woman who didn't know what she was talking about would point and say, "See? What did we tell you?" I kept my mouth shut and continued with the charade. We had to keep our hats in the ring for the corporation commission hearing that I knew would come.

We returned from the weekend to listen to a long debate initiated by Joe Farris. He stood his ground when the NRC indicated that they didn't want to hear any more about plans for dealing with "unresolved safety issues."

Questioning witnesses, Joe had established that in the previous year there had been four threatened nuclear accidents from these "unresolved issues" and that if the solutions being proposed did not work out, no available alternatives had been brought forth. "It may very well turn out that this testimony is unnecessary," he conceded, "but we don't have the information here to provide to the Board for them to find that there is reasonable assurance that will be true."

Ilene Younghein had been supplying information to the Oklahoma Attorney General's office on the Reed Report and other nuclear power problems—and about the Black Fox GE reactors in particular. The day before the hearing ended, the State of Oklahoma dropped a bombshell that tried the patience of PSO and cheered the intervenors. Attorney General Jan Eric Cartwright and Assistant Attorney General Charles Rogers told the board that because of "substantial questions about the reliability and safety of GE reactors," the State of Oklahoma wanted to participate in the hearing and to inspect a copy of the Reed Report.

Joe Farris recovered quickly from his surprise to support their request, citing the appropriate federal regulation, and was seconded by Dow Davis. Joe Gallo, of course, protested that it was not fair to do this to them at the last minute, but eventually even he agreed. Chairman Wolfe granted the petition, and the public was once again sent about its business while the board took testimony on the Reed Report.

The last day of the hearing we tried to listen carefully—or at least patiently—to discussions on how to detect and protect against equipment getting into such bad shape (through deterioration or earthquakes or any number of other causes) that loose parts would start floating around in the containment vessel and cause a meltdown. But all that day the courtroom was filled with an air of suppressed excitement.

Would a construction permit be granted? Probably. The NRC was not in the habit of denying construction permits.

Would we have to go through appeals court to stop Black Fox? Very likely.

If PSO were granted its permit, would they ask for a large rate increase to pay for the work they had already done? If they did, I was sure we had them. We could stop them at the Oklahoma Corporation Commission hearing!

All our half-fearful, half-hopeful anticipation was only setting us up for an anticlimax. Chairman Wolfe complimented the participants for their professionalism in the face of abrasive situations, promised that the board would review everything carefully before making a decision, and closed the hearing on February 28, 1979.

Far Right: The four towers at Three Mile Island, Harrisburg, Pennsylvania

Black Fox and Three Mile Island

Exactly one month after the health and safety hearing adjourned,[1] the whole country watched in horror as Pennsylvania escaped a nuclear holocaust by a margin of less than an hour. If the safety systems at the new Three Mile Island-2 reactor had been allowed to operate as designed, it is possible that the worst consequences would have been radiation releases similar to those that the NRC then considered normal. As it was, the technicians on duty (directed by an engineer who was later said to have cheated on his licensing reexamination) read some of the indicators in the control room incorrectly and shut down parts of three safety systems.

When several hours later they realized what was happening, the surrounding area had already been seriously contaminated. Fortunately, all the cooling water had not run out, and a complete meltdown was averted only a few minutes before it would have made the state of Pennsylvania uninhabitable for centuries. The slag heap it did produce was so virulently radioactive that it took nearly a decade for engineers to even begin to understand how great the danger had really been. About sixty tons of radioactive metal had melted in 5100° F temperatures and about a third of that

May Slow Black Fox on Construction Starting Date

NRC licensing board must

CASE charges PSC "has grossly misstated not only the cost but also for the revised cost esti-

had already run to the bottom of the reactor vessel.[2] In a very short time it would have started to melt its way "to China."

The whole affair had an air of surreality about it. Twelve days earlier, the release of a movie called *The China Syndrome* had provided an almost exact preview of TMI-2. People went to see the movie for that same delicious, shivery feeling we get from ghost stories: "The world is a scary place, but I'm safe!" But on March 28, we could no longer escape the realization that we needed to add, *At least for now!*

(Back in the days when children were more apt to go out to the garden and start digging holes in search of treasure than to go "malling," admiring grandparents were apt to ask if they were planning to dig "all the way to China," because we thought of China as being on the opposite side of the earth. The movie was given its name because a melted reactor core burns its way down through the layers of rock under it.)

The powerful, frightening and controversial movie (starring Jack Lemmon as the power plant engineer, Jane Fonda as a TV reporter, and Mike Douglas as the newsreel cameraman) was filmed at the Trojan plant near Portland, Oregon, with the technical assistance of our three witnesses, Dale Bridenbaugh, Richard Hubbard, and Greg Minor. An attack on greed, the movie showed the complexities and contradictions that must be faced when a California utility commits itself to nuclear power and has to choose between profits and safety.

What the movie did not anticipate, was the 1000 cubic-foot bubble of hydrogen that was formed inside the TMI-2 reactor by a chemical reaction between superheated water and the zirconium in the fuel rods. That bubble could, itself, have caused a complete meltdown. Not only is hydrogen very explosive, the bubble prevented water from getting in to cool the core. It was an enormous relief when finally, after several days, the bubble was successfully dissolved and removed. The plant was then cooled with exquisite care to prevent the formation of another bubble.

The homes of nearly a million people were invaded by radioactive fallout they could neither see nor feel. The metallic taste of the air and of food exposed to the air, however, told those who knew enough about chemistry that ions from the radioactive metal of the reactor surrounded them. Pregnant women, people with small children, farmers with dairy animals—everyone wondered what to do. No one seemed to know when or if they should leave, so—although nearly a fifth of the population decided to get out of town on its own—most people waited, trustingly. Finally, they were told to remain behind closed doors and windows. Schools were closed, and—after the worst of the radioactive discharges were over—pregnant women and young children were finally advised to evacuate the area.

Three days after the accident, CASE used some of our arduously collected funds to buy space in the *Tulsa World* for a copy of a personal appeal I had sent to PSO president R.O. Newman,[3] begging him in the light of the events at Three Mile Island to "mend the widening split within our community (and remove Black Fox's) threat to our community's health and safety."

His reply was a press conference two days later in which he told reporters that the Black Fox reactors were "simpler" than those of TMI-2. He cleverly refrained from saying, in so many words, that those simpler reactors were safer, but the implication was clear. Of course, we could not let that impression stand.

On April 11, we placed another ad[4] explaining that while the Black Fox boiling water reactors were indeed simpler—and cheaper—than the Three Mile Island pressurized water reactors, they were not safer. Instead, they were simply bottom-of-the-line reactors that, in normal operation, would release far more radioactivity to the surroundings than a pressurized water reactor. In fact, we explained, if Black Fox had had a similar accident, it "would have spilled radioactive steam into our Oklahoma countryside with the first pump failure."

In many places there is a quota of traffic deaths that must occur at a given intersection before the state will allow a stop light to be erected. When at Critical Mass '74, Dr. John Gofman told me that eventually a catastrophic accident would end utility use of nuclear power, I thought to myself, *How can they treat a traffic accident and a nuclear accident by the same standards?* Even so, when the TMI-2 accident came along, I hoped that it would be the one to stop nuclear power at last. I can never be sufficiently thankful that we were able to continue our intervention through the endless and expensive four years before TMI-2. The accident that should never have happened ultimately proved the turning point in Oklahoma's favor. If we had withdrawn, the plant would already have been half built when TMI-2 came along.

There was an immediate buzz of activity. The Oklahoma State Health Department developed an "emergency response plan" for radiological accidents at Black Fox, blithely ignoring the appeal of common sense to make such accidents impossible to begin with. State Senator Don Cummins of Broken Arrow introduced a resolution into the legislature calling for a moratorium on Black Fox construction. To gauge public sentiment,[5] a jam-packed public hearing was held in the Broken Arrow school cafeteria on the resolution, and four other state senators attended.

Most of those testifying were extremely apprehensive about living in the shadow of a nuclear power plant, especially a number of people from Inola, including Kay Ingersoll. But one man opined naïvely that "a nuclear accident would be no worse than an airplane accident or other catastrophe." I found it interesting that Senator Rodger Randle of Tulsa, was the one to tell him that "When we have a nuclear catastrophe, the effects are forever."

Tom Dalton used the opportunity to urge the committee to "seek out the confidential and controversial General Electric Reed Report and make it public." Once he had their attention, Tom took the opportunity to talk about another subject that was of great importance to him: While many of us had wondered if some of the Oklahoma Water Resources Board members had undisclosed conflicts of interest, no one had been willing to say so publicly. Tom asked the senators to investigate the board as well!

After a second hearing, this time before the Rules Committee in the State Capitol Building,[6] Senator Cummins scuttled the resolution, saying that during the course of the hearings he had switched his position and was no longer opposed to nuclear reactors.[7] Of course we were discouraged and disappointed, but I tried not to be suspicious about what changed his mind. He would have to live with his conscience if, through his actions, his friends and neighbors were endangered or killed. We took comfort in the people we reached through the publicity and discussion in the hearings.

President Carter's Commission on the Accident at Three Mile Island concluded that "fundamental changes will be necessary in ... the attitudes of the Nuclear Regulatory Commission and ... the nuclear industry." The owners of Three Mile Island tried to disclaim responsibility, suing the federal government for the NRC's "reckless disregard for the foreseeable consequences" and the reactor manufacturers for a bad design.

Testimony at the trials, however, showed that the utility company's own greed had prompted its officials to mislead the NRC in an effort to get the plant on-line before the end of 1978, thereby reaping rate increases and tax incentives amounting to $81 million. In the end, that $81 million must have looked like peanuts to them compared to what they paid out in evacuation costs and damage awards that grew out of the accident—and fourteen years of decommissioning costs. Even worse from the point of view of the utilities, the Pennsylvania Utility Commission refused to allow them to charge users for those expenses!

A sensible person would certainly argue that the NRC should not have accepted the utility's documentation and reasoning uncritically and that they did, indeed, bear the major responsibility for what happened at TMI. They had been charged, after all, with looking out for the public interest—not the public utilities' interest (although ultimately, they didn't even protect the utility). However, the NRC got off scot free when a federal judge ruled that they were immune from suit.

Even before the TMI-2 accident, General Electric—the largest nuclear equipment supplier in the United States—was losing more than $20 million a year on its reactors. In the wake of TMI-2, however, the industry's financial problems abruptly multiplied manyfold.[8] The Bank of America led the way in announcing that it would make no new loans for atomic plant construction, pending a complete investigation of Three Mile Island.

Oklahoma Attorney General Jan Eric Cartwright became very concerned about the financial viability of Black Fox in this climate, and on April 12 he asked the Oklahoma Corporation Commission to figure out the ultimate cost of Black Fox and PSO's ability to finance it. He followed this request with a motion to the Atomic Safety and Licensing

Board on April 24, requesting an indefinite postponement of the Black Fox construction permit "pending investigation of the TMI-2 accident."

Cartwright wanted to be sure that if—as the NRC had warned on March 5—hydrogen bubbles (like the one that later appeared in TMI-2) turned out to be a possible problem for all kinds of reactors, it would not be too late to cancel Black Fox Station. The warning had been issued just seven days after the close of the Black Fox hearing, but it was given greater weight by the TMI-2 accident. Cartwright clearly understood that there was no guarantee that another equally or even more dangerous discovery would be made too late to save Oklahomans from its threat.

PSO doggedly called an immediate press conference to announce that the first Black Fox plant would be completed in 1985 (instead of 1982) and the second in 1988 (instead of 1986). The delays were costing them more than $300 thousand a day, which, in three years, would add up to almost the entire original cost estimate. Rather than $450 million, they now expected Black Fox to cost nearly $2.5 *billion!* Of course, we knew that they were trying to put pressure on the NRC for a full construction permit by July 1—Three Mile Island or no Three Mile Island—and, at the same time, tell the stockholders, "It's not our fault!"

For the Health and Safety Hearing, PSO had prepared a plan for evacuating the approximately 500 people living in a two-and-a-half-mile "danger zone" around Black Fox. However, TMI-2 caused such consternation that by the following January, the NRC had increased the requirement to an area with a thirty-mile radius. That included Tulsa and Muskogee counties and three other counties, and comprised a population of about 600,000.[9] (That figure would now be more like a million or so.) In addition, to forestall people eating, for instance, beef from cattle that had grazed on radioactive pastures, (concentrating radioactivity in their meat and bones), any plant accident would now require PSO to monitor radiation in the food chain within a fifty-mile radius of the site.

Rallies, protests and demonstrations were held across the nation and in many foreign countries. Protesters against a Hanover, West Germany nuclear waste dump, seized the moment and adopted the slogan, "We all live in Pennsylvania!" On May 6, about a hundred thousand people from across the political spectrum arrived in Washington, where they assembled on the Ellipse south of the White House and then marched down Pennsylvania Avenue to the Capitol demanding "No More Harrisburgs."

Two renovated-school-bus loads of Oklahomans went to the "march on Washington" and others flew or drove. Maggie Studie went to represent the Cherokee Nation, carrying a flag in the march. My eyes were glued to the television set when Sam Lovejoy told the crowd, "Enjoy yourselves today, but don't ever forget we are dealing with a subject that is deadly serious." I sat up proudly as I saw Dr. Benjamin Spock and Lee Agnew carrying our CASE banner down Pennsylvania Avenue and cheered when Ralph Nader demanded, "Can't we find a better way to boil water?" and again when California Governor Jerry Brown called for a moratorium on nuclear energy projects.

Judging from what was visible on television, the whole thing was a lark. The reality, Nancy Perreault told me later, was somewhat different. When the television cameras were turned elsewhere, capitol police and federal marshalls behaved more like storm troopers than representatives of a proud democracy. Nancy was just one of many who were beaten up and/or arrested as they sat near the Pentagon. One policeman walked up to her and, without warning, kicked her in the stomach before handcuffing her with a group of companions and marching them off to a cold, damp tunnel beneath the Pentagon building. After a night spent shivering in the tunnel, they were carted off to jail in Alexandria, Virginia, and then released the next evening without any charges ever having been filed.

Poor Dr. Edward Teller! One had to pity him: At 71, he had spent more than half his life convincing himself that through the atomic energy he was trying to thrust down our throats he was a savior of mankind, and here we were trying to undo all his work! He took out a two-page advertisement in the July 31, *Wall Street Journal.* Its banner headline read, "I was the only victim of Three-Mile Island." The ad said that he had gone to Washington the day after the demonstration to try to undo its effects and suffered a heart attack the next day. "You might say that I was the only one whose health was affected by that reactor near Harrisburg," he reasoned, explaining that nuclear reactors were even safer than he had thought.

He just seemed pathetic to me then. But at the age of 86, he was still trying to push nuclear power down our throats: In January of 1995, he was buttonholing new congressmen and trying to persuade them that we just had to start building nukes again.

A week after his most recent foray, the Tennessee Valley Authority announced that they were abandoning the construction of three partially completed nuclear power plants because they couldn't find anyone willing to put up the money and the project was getting more expensive every day.

A week after the Washington demonstrations, the Concerned Mothers Coalition sponsored what was to be our last Rocky Point picnic before PSO closed off the park. Dr. Spock and Mary Morgan came from their home in Rogers, Arkansas. Robert, who was always happier if he could be out working in the fresh air, stayed at home to direct the strawberry picking, but he sent an enormous stainless-steel dishpan full of strawberries which we offered to everyone.

We had quilts spread out for people to sit on during the picnic. Susan Farrell had picked up an heirloom quilt with her others that morning, and Dr. Spock and Mary Morgan sat on it, eating Robert's fat, juicy strawberries fresh from the garden—until Susan realized what had happened and quickly substituted one that strawberry stains would not ruin.

In spite of our sadness over the TMI-2 accident, we spent the day enjoying the picnic—relaxing, eating, visiting, and watching the children running and laughing, as we listened to the music of our friends. The beauty of the day and the joy of sharing it with so many people who had become members of my extended family was given extra poignancy by the knowledge that in a few weeks we would no longer have the right to use the beautiful park.

In the days and years that followed, officials tried to downplay the extent of the public's exposure from TMI-2. They often compared exposure in nearby Harrisburg to dental X-ray exposure, neglecting to mention that a dental X-ray takes a fraction of a second and is confined to a very small area of the body. At TMI, people's whole bodies—including their lungs, digestive organs, and kidneys—were exposed. The strontium[90] that they breathed in and ate and drank collected in the bones and continued to radiate them continuously through their bone marrow. The clumsy cover-up attempts cost at least one honest Pennsylvania Department of Health official his job; he made the mistake of publishing research on abnormalities in newborn babies around TMI.

Bill Bryant of Oklahoma City (brother of CASE secretary Eddie Bryant) was a shining example of those who refuse to stand by silently while lies are being handed out as truth. He decided that something had to be done to counteract the disinformation that was being disseminated by people like Teller and the officials in Pennsylvania. With his brother and a friend, Jay Nelson, they put together a tabloid to showcase Christian perspectives on nuclear power and printed and distributed several thousand copies at their own expense. Its name, *The Village Square*, was a reference to Albert Einstein's admonition that it is "To the village square we must carry the facts of nuclear power. From there must come America's voice." Escher's Tower of Babel on the cover was the keynote for the whole issue, which carried speeches and articles from prominent religious leaders all over the country pointing out the depravity of condemning all those who follow us to an irreversible evil—and the virtues of acting as stewards of the earth and its resources through non-polluting energy sources and conservation.

Two years earlier, the CIA had reported a carefully hushed-up 1957 explosion in a nuclear-waste burial ground at Chelyabinsk, in the Ural Mountains of then-Soviet Russia; it had caused thousands of horribly painful deaths and made thousands of square miles totally uninhabitable. I feared it was just a matter of time before something similar happened to us, but most people who paid any attention to it at all shrugged it off as just another example of Soviet incompetence. Three Mile Island made it clear that the Soviets had no monopoly on incompetence.

Shortly after TMI-2, one of our Oklahoma Congressmen, Mike Synar, phoned me from Washington to schedule a conference. He wanted to let me know that I was all wrong about Black Fox, that a construction permit should be granted immediately, and that he would vote against each of the several nuclear moratorium bills that had been introduced into Congress. After complaining that we needed the electricity and I was standing in the way of progress, he ended on a plaintive, if unintentionally hilarious note: "Furthermore, I can't go out into my district without someone buttonholing me to tell me they don't want Black Fox!"

I gave him his conference date and assembled a couple of dozen CASE leaders. In the hour he spent with us, we calmly and kindly gave Congressman Synar a seminar on the problems of nuclear power and sent him on his way. CASE members sent him hundreds of requests for his supporting vote on the ill-fated moratorium bill. This congressman turned out to be an honest man with an open mind, and—to his cred-it—he turned 180 degrees and did vote for the moratorium.

Starting in 1977, a number of states—including two of the most populous, California and New York, as well as Hawaii, Iowa, Maine, Montana, Vermont and Wisconsin—had enacted anti-nuclear legisla-tion. After TMI-2, Connecticut passed a law requiring its Department of Environmental Protection to certify that a firm national waste disposal method existed before any new nuclear power plant could be construct-ed.[10] TMI-2 had awakened the NRC, and even though Congress failed to pass the moratorium bill, the agency imposed its own moratorium on nuclear power plant licensing, providing the time we needed to defeat Black Fox.

PSO, however, was not in a financial position to wait for the mora-torium to be lifted. On Tuesday, September 5, they finally filed their request for a 9.4% rate increase amounting to $29.4 million. W.B. Carpenter, PSO Senior vice president for finance, said they needed the money primarily because of high inflation and necessary additions to the electrical system. However, the company also asked that $180 million of construction-work-in-progress costs be included in the rate base for their Northeastern coal-fired plants at Oologah.[11]

We kept hearing rumors that some of the concrete that had been poured at the Black Fox site after the LWA was granted had developed cracks through which grass had started to grow. Eventually, Cathy Coulson-Currin and her teenaged son, Curtis, decided to see for them-selves. They rented a small plane, complete with pilot, and flew over the site. Curtis, who had been studying photography, took photographs of the cracked concrete.

This confirmed our fears that the assurances of quality control we had heard from PSO were only so much window dressing. If we could not prevent the plant from being built, we had to make sure it was as safe as possible, and if we could not trust PSO's quality control, how could we have any reason to believe that even the best designed plans would make any difference? We called a press conference to display the photographs Curtis had made.

Vaughn Conrad was sure that the cracks we saw in the photographs were actually something else. He called Cathy, our PR person, to invite me to visit the site and see for myself, and she, of course, accepted for me. The next morning, however, she received a call from another PSO employee to tell her that Vaughn had gone on vacation! "When will Carrie be able to visit the site?" she asked, astonished.

"She will not be allowed to visit the site," was the answer. "She is not welcome."

The Lure of Easy Money

In the first fifty-nine years of its existence, Public Service Company of Oklahoma had never had occasion to ask the Oklahoma Corporation Commission for an increase in the rates it charged its customers. Its rate structure had not only been perfectly adequate to cover the salaries and expenses of producing enough electricity for its Oklahoma customers but had also provided a comfortable margin for financing new facilities as they became necessary.

By 1973, however, its officers had been encouraged to believe that if they did not get on the nuclear bandwagon they would not only be out of fashion, but the federal government would not look kindly on their future activities. Washington's carrot was just as enticing as its stick was convincing: PSO's administrators had found it impossi-

CASE quizzes PSO rate

An anti-nuclear power group has asked the state Corporation Commission to consider the cost impact of the construction of Black Fox Station near Inola on electric rates, group officials said today.

Members of Citizens' Action for Safe Energy, in a letter earlier this week to the commission, asked the state panel to consider how construction of the planned nuclear power plant would affect Public Service Co. of Oklahoma's $142.2 million general rate increase request.

The CASE letter read in part " ... the financial effects of Black Fox must be faced. Furthermore, it is pure folly to believe that rate orders can be made independently of Black Fox considerations."

CASE is the legal intervenor in the Black Fox federal Nucleary Regulatory Commission licensing process.

Above: Carrie with CASE Attorney, Louis Bullock

ble to resist the lure of what was presented to them as easy money—in great quantities. PSO's strategy during its first fifty-nine years had been conservative and prudent, but in 1973, it threw caution to the winds and began a flirtation with nuclear gambling that Oklahomans are still paying for.

They hired expensive engineers to draw up designs they could use with the expensive new technology. They hired new and expensive specialized legal advisors to protect them from any unknown financial dangers of that technology. They signed contracts to buy the equipment that would support the new technology. All of this cost money—money in quantities that made their previous expenditures look like widows' mites.

Once the first step was made, it seemed impossible to draw back. If they did not continue, there was no hope that they would ever recoup their losses. And besides, hadn't other companies just raised prices to pay for their new toys? It didn't occur to them that perhaps those other companies had been in states with rampant cronyism or rubber-stamp utility boards. Oklahoma was not one of those states.

In 1973, PSO asked the Oklahoma Corporation Commission for a $9.2 million rate increase—its first request ever. Tom Dalton, acting *pro bono* for the Oklahoma Chapter of the Sierra Club, argued against it, and the OCC agreed with him. When PSO had paid its gambling debts, it no longer had enough money to continue to work on Black Fox at the same speed. It announced a delay of at least a year in the startup date.

By 1974, the company was in even worse straits. This time the request was for an increase of $20 million. Tom told the state Sierra Club board at their regular meeting that they needed to intervene, but that this time it would be more complicated and time-consuming so he could not donate his services. They would also need an expert witness, he told them, and he and the witness would both have to be paid.

The board members were reluctant to assume the responsibility for raising

David Bell, Kelley Steib, & Carrie

the necessary money. As a member of the board, I volunteered to take on the job, and the rest finally agreed. At that point I had no experience in fund-raising, just determination. But determination alone was not enough to produce the $5000 or so that the hearing cost. The fact of the matter is that Robert and I, personally, paid most of it. For several months, Mary, too, handed over her nursing home paycheck to the effort.

We felt that it was worth every penny, because the OCC again turned down PSO's request, but the price for that delay was still hard to pay. Everyone thought we were in comfortable financial circumstances, but the truth was that we all wore our clothes until they were threadbare and put off every expenditure that we could possibly delay.

It took PSO only four months to become desperate enough to go back to the OCC and ask again for a rate increase. In December, they asked for over $20 million, offering as justifications inflation and declining earnings. Both may have been true, but without the drain of Black Fox, they would not have needed the money badly enough to have risked the negative publicity their request aroused.

By the time of the OCC hearing the next spring (1975), I felt like a veteran of the rate wars. Tom Dalton, with an electrical engineering professor from OU and a California expert in the economics of rate design, represented the Sierra Club. Their main tactic—questioning the justice of charging residential consumers more and thus encouraging commercial and industrial consumers to waste energy—was not very successful. The OCC granted about three quarters of the increase PSO requested, but since we were sure PSO had not asked for a bare minimum, that could not be considered even a partial victory—especially since that fall they were granted another increase that more than made up for the cut.

Even though they had now established a precedent for receiving rate increases, it was a while before the impetus of Black Fox's cost again drove PSO to ask for a rate increase, this time for $29.4 million.[1] The limited work authorization issued in 1978 had allowed PSO to remove trees, level the ground, set up auxiliary buildings, pour concrete to stabilize the now-naked soil and sandstone, build a railroad spur and roads, and install transmission lines. All that had cost money—$139 million. And then the NRC declared a post-TMI-2 moratorium on plant construction!

Oklahoma law requires that only necessary costs be allowed in a utility rate base. Because of this, the OCC can, indirectly, restrict utility expansion to what is reasonable and necessary to meet Oklahoma's foreseeable requirements. By the time I became involved with them, PSO was no longer an Oklahoma company. Along with three others it was owned by a Dallas-based holding company. We were concerned that the motives of a Texas company would not necessarily include the good of the people of Oklahoma.

Dissatisfied with the results of the 1979 rate increase request, PSO requested the right to issue bonds to raise money. In the spring of 1980, the OCC refused to authorize bonds to finance Black Fox construction, because PSO couldn't show that Oklahoma needed the additional generating capacity.

Denied that source of funds, PSO turned to CWIP: The headlines in the *Tulsa World* on November 9, 1980, read, "PSC requests rate increase to finance Black Fox construction."

"This rate increase," admitted their Vice President of Finance, William Stratton, "is intended to sort of slide (the rate payers) into the project." Their request was the largest that had ever been filed in Oklahoma history. Politically, it was a big gamble, but without it, they confessed, they would not be able to build the plant. With it, they would have the 'shot in the arm' that would let them continue.

The OCC was being asked to approve an emergency rate increase of $59 million and a permanent increase of $136 million to cover construction work in progress on the Black Fox plant. If the OCC agreed, PSO's customers would again be asked to pay higher rates for the "cheap energy" they had been promised—even though no energy was yet forthcoming.

After Congress passed the National Energy Act of 1978, the OCC had given a grant to TU law professor Gary Allison for a legal and regulatory analysis of the rate-making standards in that act and of their applicability to Oklahoma's utility regulation. He concluded[2] that under these regulations, the OCC would have to be sure that their cost/benefit analyses of nuclear power plants included decommissioning costs. He also noted that it would be difficult, if not impossible, for a utility to build a nuclear power plant without Construction Work in Progress funds.

Tom Dalton thought we had a real chance of stopping Black Fox conclusively at the OCC hearing if we could hire "... the experts needed to do a really professionally complete job," and we wanted desperately to believe him.

We were filled with elation. The time had come, we believed, for the final reckoning, the battle that would defeat Black Fox for all time. Surely the hydra-headed monster had cornered itself too securely to slither once more out of our reach. We thought too soon.

A month later, the OCC's staff, at the urging of PSO, tried to divide the hearing into two parts. One part, the one that related to PSO's financial stability and rate design, was to be heard by a commission referee as soon as possible. That would allow the rest—the CWIP and Black Fox portion—to be put off until the following September, and by then the NRC might already have completed its hearings and granted the Black Fox construction permit.

If that had happened by the time the OCC hearing began, it would be nearly impossible for them to deny CWIP money. People would say, "They've already spent all that money; we'd better get something for it!" We had to head them off.

CASE hired Neal Talbot of Energy Systems Research Group in Boston; the Coalition for Fair Utility Rates hired an economist from Berkeley; and the Sierra Club presented husband-and-wife experts in energy and utility law, Amory and Hunter Lovins. Energy consultant Gene Tyner of Norman took on the responsibility for shepherding the Sierra Club work on the hearing, and I, of course, worked with CASE volunteers.

However, the Coalition for Fair Utility Rates decided to follow the example of big-time organizations like the United Fund and the American Cancer Society: They hired Charley Cleveland, a former state representative, at a salary of $1000 a month to raise funds to cover their expenses. Those of us in CASE wondered whether he would be able to raise enough to pay his own salary, never mind anything over to pay legal and witnesses' fees. "If CASE had paid me $1000 a month," I told them, "most of its income would have gone to me!"

Keep in mind that all this was happening before such *causes célèbres* as the PTL and United Fund scandals of the eighties and early nineties had alerted the majority of Americans to the fact that when an organization uses paid fund raisers, the amount that actually goes to the ostensible purposes of the organization is always much less than the public believes and frequently nearly nothing. We just knew what our experiences had been.

In fact, Charley was usually able to get his own salary paid, but the Coalition still owed money to its lawyer for a previous rate case when he was needed once again—and that happened during this hearing.

I sent out a letter to every CASE member, begging them to write the members of the OCC and ask them not only to hear the entire case themselves, but to keep both parts of the request together. "Tell them once again why we don't want nuclear power in our state," I suggested.

"Please remember, though," I cautioned, "the commissioners are human beings and deserve our respect and understanding: They will respond better to kindness, honesty and objectivity than to rudeness and antagonism!"

I reminded everyone about the time the commission members had come to Tulsa a year and a half earlier to listen to our opinions and viewpoints. The last three or four people to appear had unraveled most of the good work the rest of us had done by their rude, noisy, antagonistic, and disruptive behavior. I had been so shocked and humiliated, that I phoned the OCC members and apologized for their incivility.

"You attract more flies with honey than vinegar," I concluded. "If we wish to persuade the OCC members that we are right, we must not antagonize them. Let's write with dignity, and perhaps we can erase that dreadful image from their minds."

OCC chairman Hamp Baker and his colleagues, Bill Dawson and Norma Eagleton, were not only willing to listen to us, they respected our arguments. They scheduled the hearings together, and when September came, they listened to every word of the entire hearing themselves!

Tom Bomer organized a Shady Grove Concert and Picnic at Mohawk Park on Sunday, July 27, 1980, where a number of CASE musicians entertained members from all across the state. On the day of the concert there were occasional clouds and gentle, cool breezes out of the northeast. Under the canopy of tall, lush trees, everyone seemed relaxed and happy. My grandson Teddy Lemon, who was visiting from Massachusetts, helped Mary and my niece Cheri Cahalen set up for the event and then packed up and picked up before we left.

Shady Grove was so successful that Tom organized Shady Grove II for September 14. I was especially pleased to have two musicians from Claremore, Jerry Lee-Mills (whose wife Rowe chaired the Art Department at Rogers State College and had been a great help to CASE) and Michael Brewer.[3] The musicians who performed most frequently for us—Randy Crouch, Steve Curry, Phyllis Ellias, Jim and Toodles Johnson, Michael Long, Scott McIntosh, Marilyn Oldefest, Gary Steel, and Scott Williamson—had gained a following among us through the years, and many people came just to hear their "unplugged" songs.

In the meantime, Tom Dalton's efforts to reopen the EPA hearing finally bore fruit—but the fruit had a worm in it: The hearing was to be held in Dallas, making it impossible for them to be very effective from our point of view. Only those Oklahomans who could afford to leave their jobs for several days and rent a hotel room would be able to attend. Furthermore, local media coverage would be minimal. We were outraged.

At our booth at the Tulsa State Fair that fall we gathered over a thousand signatures on a petition asking that the hearing be held in Tulsa and sent it to the Dallas EPA headquarters. We copied the signatures and petition and sent them to President Carter, asking him to intercede for us. A great many CASE members also wrote to each U.S. senator and congressman from Oklahoma, asking them to use their influence with the EPA. Finally, we were told that the hearing would be held somewhere in Tulsa, but it was some time later when Adelle Harrison phoned from the Dallas EPA office to tell me that they would be held in the State Office Building.

Mary left to run an errand in Tulsa right after I told her about the call. I phoned Cathy Coulson-Currin, who was responsible for CASE's PR, and asked her to schedule a press conference for the next morning to announce the news. Not long afterward, I turned on one of the local TV stations for the noon news and was surprised to hear Congressman James Jones announcing that the EPA hearing would be held at the State Office Building in Tulsa and implying that his intercession had been the deciding factor (which it may well have been).

The phone rang, and I heard Cathy's angry voice asking how come Congressman Jones had scooped our news.

When Mary returned and heard what had happened she started laughing. It was an election year and Congressman Jones was handing out campaign flyers asking people to reelect him when Mary happened along. Mr. Jones handed her a flyer, and she handed it back, telling him she was from Claremore and couldn't vote for him. "I know someone from Claremore: Carrie Dickerson," he replied idly.

"She's my mother," responded Mary.

Pricking up his ears, he asked, "Has she heard from the Dallas EPA yet?" To Mary's surprise, when she told him about the phone call, he turned abruptly and walked quickly away. She hadn't realized that she should not have said anything yet.

It must have been hard for lawyers and expert witnesses to wait for their money. It was just as hard to bear being hounded for it—especially when, like Umesh Mathur's attorney David Madden, the bill collector was clever about finding a tender place to tread upon. "Your failure to respond to Mr. Mathur's inquiries," wrote Mr. Madden, "is remarkably like the response one would expect from some corporate giant which presumes it is unaccountable for its actions."

CASE had already paid Umesh several thousand dollars for his work in the earlier hearing, and I had already told him he would be paid for this one at the first possible moment. When potential donors asked, "Why doesn't he give his time?" I would ask in return, "Would you be able to give days and weeks of work for nothing? Umesh must make a living, too."

When we were trying to get the EPA hearing reopened, Tom Dalton told me we would not have an expert witness on water: CASE still owed Umesh $4000, and he was suing me. Since water pollution was the issue at hand and since Umesh was the best witness available on water pollution, that was unacceptable. "My shoulders are broad," I informed Tom, "but I do not carry a chip on them. I have finally managed to find the money to pay Umesh, and he will represent us!" When Umesh came to testify, I acted as if there had never been a problem. The hearing was much more important than any discord between Umesh and me.

PSO sometimes had to make do with less expert witnesses: One, under questioning by the EPA attorney, disqualified his testimony. We did not have that problem: Our attorney and expert witness had done their homework, diligently researching the issues for months in preparation for the hearing. They were more than worth their hire.

PSO contended that monitoring and measurement requirements in the permit were more stringent than the law required. Umesh Mathur countered that, "The pollution [from the Northside Tulsa sewage treatment plant and from industries] will be in the intake water and when it leaves the plant it will be concentrated tenfold at least by evaporation in the cooling towers."

At the hearing, Win and Kay Ingersoll led the Inola contingent opposing Black Fox. They had been against Black Fox from the beginning but kept their opinions to themselves "until the [Inola] Chamber of Commerce came out for it 100 percent [and] we decided someone needed to take a stand." Then they began organizing anti-Black Fox meetings—attended by as many as 200 at a time—in the 'clubhouse' on their ranch north of Inola, where several times I gave film and slide presentations.

Once the OCC had decided not to bifurcate their hearing, we hoped again that the battle would be engaged without further ado. In preparation for the hearing, the Sierra Club, the Coalition for Fair Utility Rates, and CASE united in our planning and our efforts. Tom Dalton, who was still in the process of appealing the LWA, continued to advise us, and it was partly at his instigation that in January, 1981, we hired Louis Bullock to represent us.

PSO, however, had not yet run out of potential escape routes. That March, State Representative Cal Hobson of Lexington introduced a bill[4] to allow increases in utility rates without corporation commission approval. The bill would have given the utilities a blank check, effectively removing any regulatory authority over rate increases.

Hobson admitted that utility lobbyists had helped draft the bill but tried to tell the public that consumer groups had also been involved. If there were any, they must have been ashamed of their complaisance, because none came forward to claim the "credit."

I prepared a tabloid-size newsletter to tell the public about the bill. There was no extra money in the CASE treasury, so Robert and Mary and I paid for printing and postage for 35,000 copies. CASE members addressed and mailed them to people across the state. Robert and I went to Oklahoma City twice to attend legislative committee conferences on the bill before it was finally tabled.

At the end of the first conference, Robert was standing in the hallway, discussing the bill with one of the men from PSO. "I do believe your wife could work out our differences to everyone's satisfaction," his companion told Robert.

My loyal supporter replied with a twinkle in his eye, "I could have told you that!"

The Death of a Hero

Robert Dickerson, 1959

When Robert had had his open-heart surgery in 1977, his surgeon had predicted that he would eventually suffer a stroke. That predicted doom was fulfilled the day after the second legislative conference on April 14, 1981. Although the stroke paralyzed his left side, he cooperated enthusiastically with his physical therapy and seemed to be improving. We told Patricia not to come down from Massachusetts, because we believed that, with time, he would recover.

For all of his moral and financial support in the fight against Black Fox, Robert was never able to believe that we could overcome the power of corporate giants and the inexorability of government inertia. He did not believe, however, that the probability of a negative outcome excused one from a clear civic duty. Whenever there was a meeting I needed to attend or a task that only I could carry out, Robert insisted that he would be all right without me.

The staff at Hillcrest Hospital could not have been more helpful, but Robert also received care and attention from CASE members as well as his children and grandchildren. He never lost his gallantry or his optimism and sense of humor, and at the end of a month he was again able to talk and feed himself.

On May 16, Robert suffered another stroke. By the time Patricia had flown in, he could barely mumble, and soon he could communicate only with motions of a hand and a toe—and with a blink of an eye. When we had gone to see him in the recovery room four years earlier, Robert had had tubes in nearly every orifice. He had barely been conscious. None of those things kept him from making a joke in mime. When Patty arrived there this time, he was still telling her with those few motions he could still control that the world was still a funny place and he was still laughing at it.

Those last three days of his life were more cruel to me than anything that had gone before. It was unbearable to see him suffering and trapped in a body he could no longer control. He who had always been so strong and so determined! I could only hope that the pain in his body found it as impossible to get to his brain as his brain found it to control his body. When it reached the point that no amount of determination could make his brain and body respond, I began to pray for his death with every other breath and for a miracle with the others.

Robert was, at one and the same time, a true romantic and the most practical of

A Tribute to a Gentle Warrior

BY Gene Tyner, Sierra Club, Norman

Dottie and I learned of Robert Dickerson's deat only last week—having been in Colorado Springs t welcome the birth of a new grandson at the time of his death.

We are glad that we had taken the time to visit Robert and Carrie in the hospital, a few days prior to his death. At the time, Carrie was keeping a 24 hour bedside vigil, with the help of family and friends. Later, we learned that Robert had told Car rie that he didn't want␣␣r to waste her time hanging around th␣␣␣␣␣␣␣␣ther, he wanted her to get ␣␣␣␣␣␣␣␣␣uld be alright.

␣␣␣␣␣␣␣␣␣ainst Black Fox is all ␣␣␣␣␣␣␣␣at they loved—have lov-␣␣␣␣␣␣␣␣aged their own piece of ␣␣␣␣␣␣␣␣uggle against the risk ␣␣␣␣␣␣␣ion next door. That's ␣␣␣␣␣␣mouth is--not some

Robert & Carrie with their grandchildren:

*Top: Sandra Snelling & Teddy Lemon;
Bottom (l to r):
Signe Lemon, Melissa Dickerson, &
J.J. Dickerson*

men. Years before, when our family had discussed the deaths of other loved ones, Robert had told us of his dread of being kept physically alive even though his mind was no longer able to animate his body. "Promise me you'll never do that to me," he begged us. And we agreed.

When the doctor asked me if I would allow him to insert a feeding tube into Robert's stomach, I asked, 'Will my husband ever recognize me again?" The answer was "No. He is now brain dead." I said, "If he is brain dead, there is no reason to keep his body alive. He made me promise to allow him to die with dignity." The doctor was very upset with me. But I would not allow him to do anything to prolong the life of the body from which his spirit had departed.

On May 18, Mary and I had left the room and gone walking down the hall to ease our cramped muscles. Patty bathed his face and moistened his mouth and then sat down beside him, holding his limp hand, hoping that if he could feel it at all he would know that someone who loved him was there. He gave a long sigh, and when Patty looked up, he was gone.

In forty-three years of marriage, Robert and I did not always see things alike, but we didn't allow that to affect our life together—and our marriage had been a good one. In that moment, I lost the love of my life, the father of my children, and my most ardent and faithful supporter.

In the following weeks and months, my mind had to step carefully around the void he left in my life. My grief had to be walled up in a narrow cell, until our common goal was accomplished. Now, more than ever, I had to give everything I had to the coming battle and fight Black Fox to the finish.

Robert's cousin from Missouri, Baptist minister Omer Dickerson, Jr., officiated at his funeral. Robert's funeral celebrated his life, not in the words from the lectern, but in the outpouring of love for him. Those who came to count themselves as his family and his friends included people who had known him all his life as well as those whose paths had crossed his only infrequently. CASE members who had joked with him and cared for him so tenderly when I had to leave his bedside were there. People whose parents he had looked after in the nursing home came to share some special memory with me. Men whom he had taught modern farming methods when they came home from World War II were there—and so were people from PSO!

If I had not had preparations for the final battle to occupy my time, during the next few months, I think I would have gone mad. On the farm—and especially in the house Robert had built for me—everything my eyes rested upon told me lying little stories. When I walked into a room, the dent in the chair cushion told me he had just gotten up to get a book. When I looked at the telephone it said, "He'll call in a minute. See? What did I tell you?" I would pick it up and call someone to still its lying voice.

The lies the bed told me were the worst: "Go on to sleep. He'll be home later, and he'll put his arms around you the way he always does." There was no room I could stay in, no chair I could rest in.

Victory Is Ours!

Before Robert's stroke, Tulsa University Law School had invited me to a luncheon to initiate a committee that would prepare for a September conference on nuclear power generation. Representatives of nineteen community and civic organizations were there, along with a public relations officer from PSO. When that gentleman saw me walk into the room, he protested that, "A radical group like CASE doesn't deserve to be represented."

"I can work with anybody objectively, understandingly, and without recrimination," I told him, "and I'm sure you can, too, if you choose." From that moment, he made an abrupt about face. At our meetings that spring and summer, he never failed to have a pleasant word for me. When Robert died, he was at the funeral. And at the conference itself, we always sat together.

The Oklahoma Corporation Commission, 1981
(l-r) Bill Dawson, Hamp Baker and Norma Eagleton

With the financial support of the National Science Foundation, we selected outstanding speakers on both sides of the nuclear debate from all over the country, and TU's National Energy Law and Policy Institute invited them to speak at the conference. TU law professors John Lowe and Kent Frizzell helped us present three days that went a long way toward helping average citizens understand the issues we had been grappling with for years.

The day before the Oklahoma Corporation Commission hearing was to convene, September 13, 1981, Charley Cleveland called a meeting between Coalition leaders and our attorney, Louis Bullock. One of my friends in the Coalition tipped me off to the meeting (which had not been publicized) and hinted strongly that I should attend. Because all three organizations had agreed to collaborate in the hearing, I felt CASE should be represented, so I phoned our leaders and asked them to meet me at the Coalition headquarters.

The meeting turned out to be one of the most unpleasant experiences of my life. It developed that the Coalition still owed Louis Bullock money for his work on their behalf at an earlier hearing, and Lou wanted them to clear that up before he extended them credit for the OCC hearing. There was nothing especially shocking or untoward about that. But then a bombshell exploded in our midst.

"No problem, Lou," Charley announced in a voice filled with bravado. "We won't be needing your services at this hearing, because I will represent the intervenors. Of course, we'll give you your money as soon as we can."

There ensued a barrage of words between Lou and Charley, the likes, of which, I had never before heard. In effect, Charley was firing our attorney without consulting with anybody else—certainly without consulting with CASE.

The chain of logic was inexorable: Charley, though a former state representative—and very knowledgeable about the case—was not a lawyer, and the OCC would never allow him to act as our attorney in the hearing. Without an attorney, we would not be allowed to participate. If we didn't intervene, Black Fox would be built. Lou Bullock had already spent eight months preparing for the hearing, and no one could possibly replace him.

I beckoned the CASE members out to the corridor and outlined my thoughts. When we had agreed on our course of action, I asked Tom Bomer to speak for us all. I was so distraught that I knew I would burst into tears if I tried to do it. Tom did a far better job than I would ever have been able to do. "CASE will take full responsibility for Lou's fees," he told the other people there. "We all know how important it is that we be able to reap the benefit of all his months of work."

There was stunned silence.

Charley was furious. However, eventually everyone—even Charley—accepted the justice of what we were saying: If he tried to represent us, we would not be allowed to participate. But that didn't prevent hurt feelings.

Then Tom Bomer admonished us all that, individually, we had a further part to play. "When we go into the hearing tomorrow, it must be as if the events of this day had never happened. Not one word of what we have said here today must be uttered. Instead, we must present a solid, united front to the world."

The next morning, a chartered bus and several cars stopped in front of the Jim Thorpe Building in Oklahoma City to unload dozens of people from the Tulsa area for the corporation commission meeting. During the noon recess we met for lunch a couple of blocks away at Doris Gunn's home. What a spread it was; Doris and other Oklahoma City CASE and Sierra Club members had provided a feast! Everyone had a great time, and no one would have guessed that the day before we had nearly come to blows. Lou, an old hand at this sort of thing, acted as if nothing had happened. His manner was completely urbane and professional.

Since the first day of the hearing was devoted to statements from citizens, I had an opportunity to outline the whole history of the struggle against Black Fox. I started with the first visit of the two AEC representatives, who tried to tell us that, legally, we could not stop a nuclear power plant and all we could do was suggest improvements.

I told the commissioners my reply had been that we would stop Black Fox through the simple economics of nuclear power—its hidden costs and subsidies as well as the inevitably increasing costs to the consumer. "Back then," I told them, "the AEC was telling us that nuclear power would be 'too cheap to meter.' The fact that we are having this hearing at all proves that I was right: Nuclear power is, in reality, the most expensive available source of energy."

I continued, "For eight years, I have been waiting for the day when we would sit in front of you. You are, truthfully, the only body that we can look to for an adequate understanding of the economic issues and their implications for Oklahomans. I know you have the expertise. And the kinds of questions you have asked in the past lead me to believe that you have the will to act in the best interests of all the citizens of the state."

I explained that, although the AEC and NRC hearings were little more than window dressing, our participation in them had served to make us credible participants in the current hearing. Because of our intervention, our expert witnesses had told us, Black Fox would not only be the safest and cleanest nuclear power plant ever built, it would be the last one built. "Instead," I told the commissioners, "my goal was that Black Fox would be the first nuclear power plant *not* built through the hearing process."

Although we had deserved the accusation that we were using delaying tactics, I admitted, our goal had been not delay, but denial—the denial of PSO's and the NRC's right to foist Black Fox on us and our descendants for unimaginable generations.

"I have no bone to pick with PSO's people," I continued. "Over the years, I have learned to value my friendships with many of them. I have come to think of Black Fox project manager Vaughn Conrad as another son. My own son and I don't always see eye to eye, but that doesn't make me love him any the less. I am not fighting people: I am fighting a concept, a mistaken idea. I have discovered the reality of what I was taught years ago at the little Methodist Church in Nuyaka: 'We must love the sinner and hate the·sin!'

"No one is immune from making mistakes," I concluded. "My hope is that, just as I have done so often when I discovered that I have chosen a wrong path and abandoned it for a better one, PSO will come to realize its error and turn toward avenues that will support life and health."

Finally turning to a more personal vein, I attempted to show just how important I felt it was to defeat Black Fox: "My last eight and a half years have been consumed in this struggle. My husband and I sold our nursing home and spent the entire $200,000 from its sale on the fight to stop Black Fox. I mortgaged my own land for an additional $26,500 and worked with other women making quilts to raise $60,000—all this to keep the hearings going. I could have been making money for myself all those years. Instead, I've spent all our livelihood. My husband, who had been my constant supporter and advisor in all this, died last spring, leaving me bereft. And now I must face widowhood under a mountainous load of debt—all because of Black Fox.

"My five grandchildren are almost strangers to me. My own grandmother, an untutored but gifted naturalist, came to live with us soon after my father died, when I was not yet three. It was she who opened to me the beauties of the world and who guided my small fingers into the discipline that has served me all my life—and never more so than in making those quilts to raise money. My grandchildren will have no such memories.

"Instead, when they think of me, they will remember the paper cuts from stuffing envelopes and the smell of the glue on the envelope flaps. They will remember helping me make posters for rallies and marches, attending lectures, helping operate the film projector, knocking on the doors of total strangers, and handing out leaflets at fairs. Most of all, they will remember the endless hours of waiting—waiting for me to finish trying to persuade one more person to act on his knowledge and convictions. To be sure, there are worse legacies, but that is not the one I would have chosen.

"This summer, I took an afternoon off to make yeast bread. My ten-year-old grandson, J.J., who was not quite two when Black Fox was announced, happened to come by. 'Grandmother, please teach me to make that bread,' he pleaded.

"'I will, J.J.,' I replied, 'just as soon as I stop Black Fox.'

"'Grandmother!' he exclaimed, drawing himself up to his full height and placing his hands on his hips, 'I don't want to wait till I'm grown!'

"Commissioners," I ended, "I don't want to wait till he's grown, either.

I have more to catch up with than teaching J.J. to bake bread, and I can't begin until Black Fox is canceled!"

When Ilene Younghein spoke, she took the time to tell a parallel story, including her earlier research into plutonium contamination at the Kerr-McGee Crescent facility and her later observations of the Karen Silkwood hearings. The many years of fighting nuclear power had not only absorbed too much of her life, it had affected her heart: Two months earlier she had had to have a (non-nuclear!) pacemaker inserted.

If Hamp Baker, Bill Dawson, and Norma Eagleton never did another virtuous thing in their lives, they deserve a star in each of their crowns for every day of that hearing. They sat attentively through the nine weeks that were frequently filled with mind-numbing lists of numbers, while members of our consumer groups would have to nudge each other to keep their snores from disrupting the civility of the hearing room. Lou Bullock, too, managed to stay alert throughout, despite the twelve or more hours he spent every day, five days a week, on the hearing.

However, the length of the process allowed the newspapers to tell the public the reality of what was being asked. Early on, the *Broken Arrow Daily Ledger*[1] explained that one of the major reasons for the size of PSO's request was that rather than a reasonable, just rate of return, they wanted to be able to earn more than 16%—more than the rate of return on money market accounts at that time. To come up with all that money, they wanted to impose a nearly 40% increase in residential electric rates. Lou told the commission that if PSO's figures were extrapolated into the future and the costs of building Black Fox were included, indications were that residential customers would eventually have to pay increases of from 150% to 300%!

Nancy Strainer, PSO's manager of economic programs, maintained that the increase was justified because "peak demand for electricity will leap 45% in the next decade."

Lou retorted that "... continued utility rate hikes will cause consumer interest in conservation to snowball."

While Nancy Strainer admitted that price increases would certainly encourage conservation, she persisted in maintaining that "a growing 'conservation ethic' will not have enough impact to raise serious questions about PSO's construction plan."

At the hearing, Neal Talbot explained the deficiencies in PSO's request for CWIP. Amory and Hunter Lovins talked about their 'soft path' energy concepts and later they met personally with some of PSO's people for an in-depth consultation on the use of renewable energy, conservation and energy efficiency.

PSO people were far from stupid, and the years of reading about and discussing other companies' disastrous flirtations with nuclear energy must have given them pause—even as they publicly maintained their complete dedication to Black Fox. During the time the hearing was in session, newspapers were reporting that thirty-eight plant proposals had been abandoned in the previous two and a half years. It was also at about this time that the Diablo Canyon plant was permanently mothballed, without ever having produced a single ampere, because it had been built on a major fault line.

At one point we saw a tiny crack in the PSO facade when their vice-president of corporate planning and performance, Frank J. Meyer, hinted that they were "not wedded to" Black Fox. That hint was given additional weight when their own consultants concluded that Black Fox would cost nearly four times the price of a coal-fired plant of similar capacity. (While I am no advocate of coal-fired plants, at least the damage they cause is reversible in decades rather than millennia!)

When the hearing ended on November 10, we had no inkling of what the Commission's eventual decision would be, but there was hope in our hearts. With renewed energy, we set out to find the money to pay off our debts. Periodically over the years, I had sent funding requests on behalf of CASE to every foundation that had an interest in problems that might be related to the risks of nuclear power. I also sent copies to the national anti-nuclear organizations and asked them to help me find funding. The responses from the foundations were invariable and disheartening: They did not fund legal intervenors, only civil disobedience groups.

However, two weeks after the hearing adjourned, there was a new answer: A foundation that wished its name not to be released sent me a letter to say that a $20,000 check designated for CASE's Black Fox work was being mailed to the Sierra Club Foundation. Even now, I can relive in memory every shred of emotion—beginning with resignation that this was going to be another form letter, passing through cautious hope, incredulity, and, finally, joyful relief! No money ever arrived more opportunely.

I was later to discover that one of my letters had been lying on the desk of one of the people I had met in Washington at a Critical Mass conference. (After all this time, I am no longer certain whether it was Anna Gyorgi, or someone at NIRS, or another person entirely.) A staff member for one of the foundations I had solicited so unsuccessfully, came in to say that they had some unallocated money, and to ask if anyone knew of a particularly pressing case. He was handed my letter.)

We triumphantly handed Lou Bullock a check for $10,000 of that donation. Another check for the same amount was sent to our expert witness, Neal Talbot, in Boston. We all had a breathing space. We could take the time to relax and enjoy the holiday season. But I still could not give myself permission to mourn Robert's death. I could not let down my guard until Black Fox was defeated.

After Christmas came three more anxious weeks before the Corporation Commission released its decision. Finally, on January 15, 1982, our happy, incredulous eyes read, "PSO, Western Farmers Electric Cooperative, and Associated Electric Cooperative should not proceed with this project."

Firmly declaring that it "can and will continue to protect Oklahoma rate payers from imprudent management decisions," the Commission continued, "... the Black Fox Nuclear Power Station project is no longer economically viable ... Construction Work in Progress for this project should not be allowed in Applicant's rate base ..."

PSO was in a cleft stick. GE had letters of intent from them contracting for two mind-bogglingly expensive reactors; two electric cooperatives had paid them $163 million for forty percent of Black Fox; and they had signed firm contracts with these same cooperatives pledging to build and manage it. But they did not have the financing to complete their contracts. And they had thirty days to decide what to do.

For a month we hardly dared to breathe. Would PSO throw in the towel, or would we suffer another defeat?

Nowadays, we have been treated so often to the spectacle of the CEO being given a golden handshake to take the fall for a company misjudgment, that we have come to recognize that it heralds a serious about-face.

We did wonder what was going on when Martin Fate became the new President of PSO, yet his appointment did not give us any confidence that we had won. That came only with the announcement on February 16, 1982: PSO would scrap Black Fox!

I had known Norma Eagleton and her family, the Haddads, most of her life. As a child in Claremore, she was already known for her incisive intelligence and her beauty. It was no surprise to anyone here when she went on to become a lawyer and to gain respect—and the kind of influence that comes from respect—in Tulsa and, later, throughout the state. While we had never been more than acquaintances, we had always had a certain esteem for each other. Thus, when the Commission was notified formally of PSO's decision, it must have seemed natural to her to call and tell me the wonderful ... joyous ... shattering news.

As I picked up the telephone, there were several reporters and television crews, waiting to record my reaction—whatever the decision. When the telephone rang, they all came to attention in the blink of an eye. The result must have been written large across my face. Not to mention my body. I felt like I was exploding. My body couldn't decide whether to laugh or cry or scream. I wanted to jump up and down and dance. Then the other phone was ringing.

It was Bob Mycue from the *Tulsa World* to tell me the same glad tidings! I sat there, incoherent with excitement, a phone at each ear, and just grinned from ear to ear.

For nine years I had tried to hold on to a forlorn hope. For the past month, my every waking thought had been, "Will they or won't they?" And now we had won. What a wonderful feeling! I was overwhelmed with gratitude and thanksgiving. Victory was ours!

The next morning, Bob's story quoted Martin Fate saying that CASE and the other local intervenors, through their delaying tactics, were a major factor in Black Fox's downfall. "When you have an unprotested hearing," Fate said, somewhat ruefully, "it does not take long to arrive at the licensing stage."

The other major factors in PSO's decision had been the cooperation of their partners and the OCC's promise of a financial bailout if they decided to terminate the project.

After nine years of feeling like a Cassandra every time I voiced my confidence in our eventual success, I was vindicated. Of course, I had not known that inflation would become our ally, but I had said we would stop Black Fox in the OCC hearing when PSO asked for CWIP and that we would prove Black Fox too expensive for people to buy the electricity. And that is exactly how Black Fox was stopped!

The Victory Celebration

One of the best things about that day was sharing the news with many of the people who had worked so long and so hard to bring about this happy fulfillment. Ilene Younghein was, of course, the first person I wanted to call. However, in the wake of her recuperation from having a pacemaker implanted, her husband, Gaylord, had taken her on a six-month trek through the South Pacific, so I thought I would have to call out the FBI to help track her down!

When I finally reached her by phone in New Zealand, she and Gay had just come back from a climb on a Mount Cook glacier. Ilene had nearly given up on the possibility of Black Fox ever being canceled, so it took a while for my news to sink in. When she and Gaylord finally believed it, she let out a shriek that I could almost hear without benefit of the telephone!

I still treasure the letter I received from our co-intervenors Lawrence and Lottie Burrell after my phone call to them:

> Dear Friend Carrie,
>   It was sweet of you to think of us and want to call us last night. We do rejoice with you

Anti-Nuke Activist To Receive Award

...2 year-old former reactors, must be made   Fox nuclear power
...tered   public.   plant is located 12
...PSO's   Black Fox   miles north of Mrs.
...built in   Dickerson's farm near
Claremore.
...ckerson said

in the victory you (and the Lord) have won. We know you will not forget to give Him credit for His many blessings.

Allow me to say that I would certainly not wish to be placed in a position where I would be obliged to face you as an "enemy." I mean that as the greatest compliment of which I can conceive.

We trust you have been able to save the farm for yourself.

With every good wish for the Lord's blessings and with our love to you.

Lawrence and Lottie Burrell

After all those years of holding up my spirits in the face of all the people who said it couldn't be done—that I was putting my (and their) money down a rat hole—I needed to hear every one of those words.

The editor of our local newspaper made sure I didn't get a swelled head. His editorial on the Black Fox cancellation concluded, "And future generations will say, 'Carrie Dickerson caused them to freeze in the dark!' "

The only thing missing was the sight of Robert's face. How glad I was that he had been there to share in the pride and pleasure when two years earlier the Sierra Club honored my efforts by establishing a special service award called the Carrie Dickerson Award, when KRMG gave me its November 14, 1980, *Pat on the Back* for "… a fine example for the citizens of the City of Tulsa and the State of Oklahoma," and especially when two days later the Community of John XXIII had, in an elaborate and impressive ceremony, presented me with its Nonviolent Action Award.

When I heard William Nerin say, "Mrs. Dickerson has won the respect and admiration even of those who oppose her views on nuclear power," I knew I had succeeded in acting on the principles Robert and I both believed in. Surrounded by family, friends, and well-wishers, I had been given the impetus that was to carry me through the dark year that followed. Robert had not lived to see the triumph, but he had shared in the blessings of that love and faith.

Throughout the nine years of the Black Fox proceedings, I was often the target for media chastisement. (I suspect that it had more to do with publishers trying to curry favor with big advertisers like PSO than with what I was doing.) Now, suddenly, people could not say enough good things about me. Reporters interviewed me so often that the Tulsa Professional Chapter of Women in Communication named me one of their three "Newsmakers" of the year. The Claremore chapter of the American Bar Association gave me their "Citizen of the Year" award because I had worked with—instead of against—the system. The Giraffe Project, based in Washington State, honored me with their Giraffe award because I had been willing to "stick my neck out." Larry Bogart, from Citizens' Energy Council and Lois Gibbs with the Citizens' Clearinghouse for Hazardous Wastes, each bestowed their special awards of appreciation to me. Julia Frasier, among others, nominated me to the Oklahoma Hall of Fame.

I consider that all these awards were really a tribute to the many thousands of people across Oklahoma and the nation who joined in our effort to stop Black Fox.

The Sunday after the cancellation, about 200 of us gathered for a victory celebration at Owens Park in Tulsa. Oklahoma Corporation Commission members Norma Eagleton and Jim Townsend came to celebrate with us. Bill Dawson, in ill health, had resigned from the OCC, and Jim had been appointed to fill the vacancy. In Jim's speech, he described me as "a great lady whose name will be not only in the history books of Oklahoma, but in the history of the nation." I appreciated his kind words, and I was especially gratified when Norma Eagleton expressed her appreciation of the way CASE "kept everyone calm, so we could do our work."

One of the people who came to our celebration was a research chemist who had recently moved to Bartlesville from "up north." He was unwilling to tell me his name, but he said he had helped write the contentions that Mary Sinclair had sent me. Those contentions, used in fighting the Midland, Michigan, plant, had been the original basis for the ones we used in the Black Fox hearings. "Now the OCC findings may be applied to the Michigan plant," he rejoiced. "We have come full circle!" (Mary recently told me that the plant had been built but had never had a reactor installed.)

People like David Mitchell and his wife, Belle, who had been faithful contributors for years but whom I had never met, came to the celebration. (Even today, people are still coming up to me to hug me and say, "I helped you stop Black Fox!")

298

One after another, people came up to me all that day to give me gifts and tokens. It felt strange, but sweet, to act as the focal point for their emotion. The whole day was filled with a tremendous outpouring of relief and joy. "The Black Fox intervention is one of the nicest experiences I've had in Tulsa!" exclaimed one. Others expressed their relief that Black Fox was no longer casting a shadow over their lives.

I reminded them that Black Fox would not have been stopped without our attorneys—Tom Dalton, Joe Farris and his colleagues, and Louis Bullock—our many expert witnesses, and the corporation commission and told them how much we appreciated their efforts. I thanked PSO and the electric cooperatives for acknowledging the futility of continuing the pursuit of Black Fox and for turning away from nuclear power. Then I told the crowd how thankful I was for the work of every one of the thousands of people who participated in the effort, including members of the Sierra Club, the Sunbelt Alliance, CARE, CABF and CFUR, CASE, and the Oklahoma Wildlife Federation. I did my best to tell everyone what each of them had meant to me for all those nine years, but nothing I could say would have come anywhere near the reality of my feelings. It would have taken days and weeks to tell everyone how much I appreciated the things they had done for me, personally, to make my load lighter, how Dennis and Susan Farrell had kept my car in tip top running order as their contribution to the fight, how Eddie Bryant had volunteered his time and energy to keeping the CASE office in order for so many months, how ... There is no end to the list.

During all the years of the hearings, Tom Dalton had worn ordinary business suits instead of the expensive pin-stripes of the PSO and NRC attorneys. With his business suits, he wore the same beautiful necktie to every hearing, and I had admired it all those years. By the time Black Fox was canceled, his tie was threadbare and gravy stained, but still the most beautiful necktie I had ever seen. That day I finally worked up the nerve to ask diffidently, "Tom, when you get ready to discard that tie, will you please give it to me?" He removed it then and there and handed it to me. It is one of my most prized possessions!

The next issue of the *Oklahoma Sierran* featured an editorial by Kelly Jennings, "Victory at Last ... Black Fox is Dead!" in which she thanked Tom

Dalton, Joe Farris and Lou Bullock for their advice, time, and services and me for leading the fight.

> We particularly thank her for the type of leadership she exemplifies. She focused on the issues rather than the personalities involved. She always encouraged those of us working with her to treat with respect those who opposed our position. Further, she taught us how citizens can truly participate in a democratic society. She ... fought big government and a large corporation and won. And she won with class.[1]

The cancellation, however, did not cancel CASE's financial obligations or relieve my worries. In spite of the $20,000 grant we had received before Christmas, CASE still owed our attorneys and expert witnesses over $65,000. We also owed money for loans, for printing, and for other odds and ends. Five years earlier, a woman in Collinsville had lent CASE $5000 so we could start a hearing. All those years, we had been able to pay her only a small amount of the interest. We started a fund-raising drive to clean up all those debts. People did give us some money, but it is difficult to raise money after the fact, and a common reply to our requests was, "I thought Black Fox was canceled!"

In their rate order, the OCC had urged PSO to pay the fees for Lou Bullock and CASE's expert witness. This had been a pleasant surprise— it was a first in Oklahoma—but since they had "urged," rather than ordered that relief, we really didn't expect anything to come of it.

Lou Bullock and one of PSO's attorneys argued back and forth over the issue for weeks without coming to any agreement. Finally PSO's Senior Vice President of Finance, Will Stratton, set up a meeting with Lou and me. He began by complimenting CASE on our attorney and witness. Then he became very business-like. "Because they did a truly excellent job and avoided acrimony, PSO will pay the total attorney fee at the same rate as we pay our own attorneys, $100 an hour!"

We were dumbfounded. Lou and our other attorneys had been charging us only $60 an hour. What a wonderful bonus for their labor of love! In all, PSO was offering to pay over $86,000 to Lou and Neal.

As we prepared to leave, Mr. Stratton asked me to come by his office the next morning, to pick up the two checks. "I'd feel very unwelcome in your big, downtown office building," I demurred.

"You'd be surprised," he urged me. "There are a lot of people in that building who feel we owe you a debt of gratitude!" That was all he said, but in the context of the rumors that were going around Tulsa that Black Fox was bankrupting the company, he probably didn't need to continue.

The next morning, Vaughn Conrad greeted me at the door with a big hug. I treasure the memory of that moment, because I have never seen him since. I often wonder where he is now and hope that his life is a good one.

Will Stratton handed me a package containing a little T-shirt with a stylized black fox and a 'Pro Black Fox' bumper sticker! We laughed over his mementos and said good-bye. He left Tulsa and went back to the parent office in Dallas.

When I delivered Lou's check to him, he thanked me, explaining that had I not treated PSO's people with kindness and understanding and brotherly love during the whole hearing process, PSO would most probably have ignored the OCC's suggestion.

When they received the checks from PSO, Lou and Mr. Talbot sent back the $10,000 each we had given them earlier, and Lou also returned the extra several thousand dollars we had paid him over several months. Those refunds, along with what we had been able to raise during our post-cancellation fund drive allowed us to finish paying off our other attorneys and witnesses and to pay some of our other debts, like the one to the woman in Collinsville.

Unfortunately, one of the newspaper articles said that I had been the recipient of the money PSO had paid Lou and Mr. Talbot. Many people read the article and believed that, as they put it, I was "rolling in dough," when the truth was that I was in perilous financial straits.

Many of the people who had so desperately wanted Black Fox stopped were angry at the rate increase the OCC had granted to permit PSO to pay for the expenses it had already incurred. Instead, they believed that the stockholders should have paid the cost, and I received many angry phone calls from PSO customers demanding to know why I had "allowed" the OCC to require them to pay $240 million over ten years.

I sympathize with their point of view, because I, too, believe that if stockholders were held accountable for the foolishness done in their names, they would soon demand an accounting that would make businesses behave much more responsibly. However, there was only so much I could do. Instead, I patiently asked each person, "Did you hire an attorney and witness to address this issue at the OCC hearing?"

"Of course not!" would be the answer. "I'm just a private citizen."

"I did, and I'm just a private citizen," I would tell them, "and it took every cent my family and I had worked for over many years—more than $200,000. It took the $60,000 I helped raise making quilts. It took all my time, night and day, for nine years—time for which I had neither asked nor received compensation and during which I could have been earning a comfortable living. I have spent all the money that was to provide for my old age, and I am now deeply in debt.

"Where I spent an average of $50 a day for nine years," I would continue, "you are being asked to pay about seven cents a day for ten years. Instead, you and the other PSO customers could be paying billions in increased rates because of Black Fox—not to mention additional billions to deal with radioactive wastes and damage to the environment and the health of present and unborn generations."

This attitude cost me many friends. Others, however, were more understanding. Several years later PSO maneuvered itself into a position that allowed it to refund part of the money customers had paid. Claremore residents Ann and Ward Emrick and Mary McAnally, members of Tulsa's College Hill Presbyterian Church, among others, proposed via newspaper and radio that "area residents who wish[ed] to thank me for my perseverance" send their refunds to me to repay some of the money Robert and I had lent to CASE. It was balm to my heart.

And I got my "fifteen seconds of fame" in 1989, when *People* magazine named me one of its ten people who "… Did the Right Thing"[2] during the Eighties.

The Aftermath

When the Health and Safety Hearing ended, Joe Farris had come to me and apologized: "Carrie, we're sorry we ever asked you to mortgage your land. You were right to make us go on; we may actually be able to stop Black Fox. I'm sorry for the anguish we caused you." What a relief it had been to know that I would not have to betray my sister. And what a relief it was when we finally paid Joe all we owed his firm! They released the mortgage on the Newcastle farm, and finally, apologetically, I told Clara what I had done. As I had trusted, Clara understood and forgave me.

By the time Black Fox was canceled, Clara had retired from full-time teaching, and she, too, was finding it difficult trying to manage on her pension and what little money she earned from teaching in-service classes in the Portland schools. We decided to subdivide the Newcastle farm and sell the land, assuming the mortgages ourselves. Eddie Bryant bought the first plat. If I hadn't been so deeply in debt, I would have given it to him out of gratitude. It took a long time to sell all the land, and even when all the plats were sold, the net proceeds divided by two was not nearly enough to cover all my monthly commitments. But it made the difference between desperation and outright bankruptcy.

Grateful for our victory, I now felt a responsibility to encourage others in their efforts to stop other nuclear power plants. For several weeks, Marjorie Speas drove me to rallies in Kansas, Missouri, and Arkansas to spread the word that legal intervenors can succeed.

Aside from that, however, I no longer had a dozen duties to fill every waking moment, and I could relax. As soon as I did, my grief for Robert overwhelmed me. I was sixty-five years old, exhausted from nine years of fighting Black Fox, and deeply in debt with no money in the bank. I had to give some serious thought to how we would all survive economically, but there were many days when just getting up in the morning seemed an insurmountable task. When I was at my lowest ebb, a group of friends from CASE pooled their money and bought me a round-trip plane ticket so I could visit Clara at her home in Portland, Oregon.

I stopped off in Montana to visit Clara's daughter, Carrie Mae, and her family, and standing in Carrie Mae's doorway and looking out at snow on the mountains in August, I drew my first real breath of freedom since that May morning in 1973. Clara and her husband Marty joined us in Libby and drove me back to Portland through the white ash residue from the 1980 Mt. St. Helens volcanic eruption. Clara told me of all the exciting things that had been going on in her life. She was especially proud of an award she had received from her colleagues for her "creative and humane" teaching. They had asked Senator Mark Hatfield to fly a U.S. flag over the Capitol Building so they could present it to her with the award. My nephew Rod came down from Alaska, and my older brother, Joe, came from Eugene with his wife, Irene, to take me on a tour of Oregon and the redwoods of Northern California.

If I had not had a farm and family to return to, I do believe I would have stayed with Clara in Oregon. The trip helped turn my life around. When I returned, it was with renewed energy to take up the struggle of making a living in my old age.

Because all our money had gone to stop Black Fox, Robert had decided to start a berry farm—which needed irrigation. He mortgaged his parents' house to set up the irrigation system, but he died while it was still incomplete. He had other mortgages that had to be kept up, as well. The farm had been lying fallow since Robert's death, and I had neither the heart nor the expertise to make it a paying proposition. Even living in the house where I had dwelt so long with Robert and where every board and nail spoke of him had become unbearable. I knew I had to make a radical change in my life, but at my age what could I possibly do?

Several years earlier, the Presbyterian Church in Claremore had auctioned off its parish house to make room for an extension on their church building. Thinking that Patricia might move with her family to Oklahoma and live in the parish house, Robert bought it, moved it to a beautiful spot near the canyon on the farm, and started to build a stone heat sink in front of the house; he planned eventually to add a solar collector to heat the whole house. But Patty stayed in Massachusetts after all, and an idea of turning the house into a daycare center for Mary to operate fell through when I commandeered all her time for Black Fox. The big house remained empty.

I was still a registered nurse, and I still had experience caring for older people. Perhaps the solution to my problems lay there, I thought. After considerable reflection—and several rejections—I managed to get a bank loan to renovate the parish house. Mary and I had five bedrooms partitioned off the enormous open space and bathrooms and a kitchen put in so we could live there and operate it as a boarding home for older people. By the time we had made Aunt Carrie's Country Cottage livable, however, I had also had to mortgage the house that Robert had built for me.

Robert and I had had a phobia of mortgages since 1943, when I had to leave my two babies to teach school to pay off his father's mortgage on our farm.

With the help of our children, he had added rooms and conveniences to our house, piecemeal, as we saved enough money ahead. Once the farm was paid off, we were determined never to leave ourselves at the mercy of a bank again. Convincing Robert and his father to mortgage the farm to build the nursing home had been a nearly impossible task. And now that I was buried under mortgages, I no longer had Robert to help me figure out how to manage.

Our three expert witnesses, Greg Minor, Dale Bridenbaugh and Richard Hubbard, sent me tickets to take J.J. to Disney World in Florida. At the time, Irene Dickinson and her husband Dick lived just forty miles from Disney World, and they invited us to stay in their home. So while the parish house was being renovated, J.J. and I took off. Irene and Dick not only took us to visit Disney World but shepherded our explorations of several other areas in Florida. That week with J.J. and Irene and Dick is one of my most cherished memories.

Caring for our boarders was hard work, but it also held many rewards for Mary and me. Our old friend Daisy Diffie spent most of the last few of her almost 100 years of life with us. CASE supporter Armin Saeger brought his father, Armin, Sr., to live with us. A gentle, retired teacher with a stimulating mind, Armin, Sr., lived to be almost 96. While he was with us, the Tulsa Society of Friends came once a month to hold services followed by covered dish dinners. Over the years, I had become well acquainted with his grandchildren—especially with Bob Saeger and Bob's wife, Marla; they became tireless workers in Oklahoma's defense against nuclear facilities and nuclear waste.

Eventually, the work of the boarding home became too much for me, and Mary could not do it alone, so we found other places for the people who were with us and began looking for other ways of supporting ourselves. I began sharing Grandmother Perry's hard won lessons with new generations of quilters, and Mary and I began a small business selling earth-friendly products and herbs.

I am often asked if I regret spending all that money on the battle against Black Fox. One answer is that of course I wish others had put more money into the fight so Robert and I would not have had to sacrifice everything we had spent our lives working for. Another answer, though, is that I am grateful to have had the money to sacrifice. Perhaps in God's plan, one of the purposes of the nursing home was to help save us all from Black Fox!

Although I have lost touch with many of the people I learned to love so much during our work at CASE and others have moved far away, many of my dear friends still call and come to visit and to work with me on problems that we all face today. Almost every year since the Black Fox cancellation, they have had a get-together on my birthday.

Norma Eagleton once used the Black Fox experience as an example of the importance of public give and take and intellectual honesty in devising satisfactory solutions to civic problems. In her 1987 speech to the MidAmerica Regulatory Commission, she talked about her own reasons for entering government work: "People in government," she said, "seemed to know everything about everything. They seemed so wise. What did I discover? We're only people groping for answers and solutions to problems just like everyone else."

Too many of us, because we have never participated over the long haul in public debate and have never held government office, assume that by virtue of their positions public officials have all the answers. For our government to continue to work, we must stop indulging ourselves in this kind of intellectual and moral sloth. We must take responsibility for our own democracy, or soon we will cease to have one. We will then be at the complete mercy of those whose only motive is greed.

There is no reason to give up and lose all our ancestors suffered and died for—or give up the future of humanity and, indeed, all life on this planet. What it takes is a commitment to personal responsibility from each of us. With that commitment, we can fight city hall and win. It may involve difficulties and hardships, but if the goal is worthwhile, those difficulties and hardships are a small price to pay.

Black Fox showed the way. It was a training ground for those of us who were involved and is an example for the generations that will follow us. The proof lies in the continued triumphant vigilance of those of us who went through that trial by fire and in the new generation of people committed to the future that has arisen from its ashes.

Epilogue

Keeping House

The commercials for cleaning products used to urge housewives to enlist the company's White Knight or genii in the continuing campaign against dirt and dirt-borne disease. During the years when we were struggling to stop Black Fox, I thought of what we were doing as a battle against ignorance and greed. In recent years, I have come to realize that the housekeeping metaphor fits better, because housekeeping is a daily responsibility.

No one is born knowing how to keep house, but those who have learned to do it well know that a task postponed is a task made more difficult. In our battle against Black Fox, we learned the skills to keep house in a political and public-opinion context. We learned that every day is a new challenge, that in taking satisfaction in each day's accomplishments we give ourselves strength to wake up the next morning and go back into the fray. We also learned that shared tasks are lighter ones.

Jessie DeerinWater was the first of us to apply what we had learned to a new housekeeping task.

Several years ago, a group of about a hundred Indians came walking through Claremore on their way from California to Washington to air their grievances to Congress. I felt that Claremore, as a part of the Cherokee Nation and the home of many Indians of many tribes should welcome and feed them, so I talked to people of Indian descent and various other people around town. No one seemed to have the time to do anything until I called Jessie Fink. Together we cooked all day, and by that evening we had huge dishpans full of fried chicken, corn on the cob, green beans, potato salad, cole slaw, fresh-baked bread and watermelon to set out on the tables at the little park across from the Indian Hospital.

During the latter years of our struggle against Black Fox, Jessie, a Claremore woman of Cherokee ancestry, was one of the people I could always depend on when I needed help, but by 1984 she had remarried— her new name was DeerinWater—and moved to Vian, a community near Gore in far eastern Oklahoma. Gore, Oklahoma, was where Kerr-McGee had built its Sequoyah Fuels uranium processing plant in 1970—seven years after the death of one of the company's founders, Oklahoma's beloved Senator Robert S. Kerr.

When Jessie learned that Kerr-McGee was planning to inject dissolved wastes from the plant—wastes of nearly unimaginable toxicity—deep into a fault line, Jessie's Black Fox experience meant she already knew what to do. She immediately organized Native Americans for a Clean Environment. She mobilized citizens of the community and state so effectively that Kerr McGee abandoned their injection well plans in 1985.

But the housekeeping tasks did not disappear with that victory. The wastes in question were 21 highly toxic heavy metals[1]—three of them, radium[226], thorium[230], and thorium[232], strongly radioactive—in a solution of nitric acid. They are byproducts of the conversion of uranium oxide to uranium hexafluoride. At that time, the waste solution was being treated with barium salts to reduce its radioactivity and was held in enormous ponds to await permanent storage. This makeshift arrangement could not continue indefinitely.

Not allowed to inject the wastes into the fault, Kerr-McGee cast about for a new tactic. They found it. No one could possibly object to adding fertilizer to the environment, they reasoned, so all they had to do was make the wastes conform to some government definition of fertilizer, and all their problems would be over!

They added enough ammonium to the waste solution to give it a nitrogen content high enough to give the NRC an excuse to approve a fertilizer label for it. Then they spread it—as raffinate *(q.v.)* fertilizer—on 10,000 acres of pastureland they owned. They then pastured cattle on the land and sold the beef to the public. Wildlife on the contaminated land developed mutations and genetic defects.

These wastes were only part of the contaminants produced at Gore. From the date of its opening in March, 1970, uranium escaped into the atmosphere at levels higher than even the NRC limits. Plant accidents released more in the form of UF_6, while uranium-containing solutions that spilled and leaked were allowed to drain into the Illinois River at its confluence with the Arkansas. One such leakage involved 32,000 pounds of UF_6!

One accident lead to the death of an employee, James "Chief" Harrison, and the exposure of over a hundred others to a cloud of corrosive contaminants. In August, 1990, eight other workers were exposed, provoking great public outcry.

Residents living near the plant site are plagued with respiratory problems, cancer, and leukemia at rates statistically far higher than normal. In
one family—that of Wanda Jo Kelley—13 of its formerly healthy members
died of contamination-induced illnesses. Before her untimely death from
cancer, Wanda Jo herself became another of the housekeepers working to
document these problems.

Robert S. Kerr was an American statesman who loved Oklahoma and
its citizens. His book *Land, Wood, and Water*, from which my father-in-law
often quoted, illuminates the mind of a man of conscience and integrity. I
am convinced that had he lived to see what his company proposed to do at
this plant and at others that they built, these housekeeping tasks would
have been much lighter. While Senator Kerr was not immune to the lure of
wealth and he might not have had enough clout to control the company
absolutely, I firmly believe that when he realized the consequences of its
actions, he would have done his level best to move it in a more positive
direction.

Eventually, Kerr McGee sold the plant to General Atomics. In 1986,
CASE joined NACE in an unsuccessful attempt to prevent General
Atomics from constructing a second facility at Gore. This facility used
depleted uranium to produce the material for military munitions.
Depleted uranium is a misnomer: It is not usable for chain reactions, but it
is still highly radioactive. Armor-piercing bullets, whose tips were manufactured from the depleted uranium, were used in the 1991 Persian Gulf
War to fire upon tanks on Kuwait soil. The tanks were plated with depleted uranium. When the bullets struck the tanks, the uranium melted at a
very high temperature, causing everything in the tanks to burn and melt,
including the enemy soldiers. At the same time the melting of the uranium
caused the release of uranium oxide into the air whence it was free to blow
in the wind and to settle on the sand. Or flow in the water. The British
Atomic Energy Commission estimated that the particles of depleted uranium left from the Persian Gulf campaign will eventually cause 50,000 cancer deaths. One wonders if our Desert Storm veterans' medical problems
may have been partly a result of inhaling the uranium oxide.

When Jessie decided to go to law school, Lance Hughes, who
worked with a Cherokee tribal agency in Tahlequah, took over direction
of NACE. Kathy Carter-White, a native of Tahlequah and a law student
at TU when we were fighting Black Fox, became another of the moving

forces in the organization. (I always felt that Kathy was a spiritual relative, because, she often came to our meetings barefoot—while I often kicked off my shoes during meetings.)

Through the Freedom of Information Act, NACE discovered that soil on the site had become completely saturated with uranium. Lance did a superb job of publicizing what they had found out, and in October, 1991, the plant was shut down.

When SFC threatened to reopen the plant in March, 1992, there was another housekeeper ready to help: The Cherokee Nation collaborated with NACE to file a restraining order. Wilma Mankiller, then the Cherokee Nation's Principal Chief, told the court, "We do not believe this facility can be operated safely."

In an effort to attract attention to the problems there, Bob Saeger pitched his tent across the road from the facility and went on a hunger strike. He told interviewers, "There will be other jobs, cleaner jobs." He was soon joined by people from a number of organizations, who formed a protesters' tent city. Every weekend for many weeks Bob and his fellow housekeepers continued their sit-in vigils.

Finally, in February, 1993, eight years after Jessie had started the NACE campaign, the uranium hexafluoride facility closed for good, and four months later the depleted uranium facility followed suit. The company estimates that it will take 10 years and $21 million to remove the contaminants from the site. Many believe that complete decontamination is impossible.

Lance continues to monitor the cleanup and document the long-term effects of the disastrous events at Gore. He is determined that people will know what really happens when they are subjected to these kinds of radiation and these toxins. Anecdotal evidence may point the way, but only solid records maintained over considerable periods can prove or disprove conjectures about long-term effects. Lance does us all a service in adding to the growing body of meaningful evidence—and in making sure that the evidence is not buried or discredited by unrebutted or ineffectively rebutted disinformation.

All the people I have worked with have done their part, and all over the country and the world, there are others who are making themselves responsible for these tasks. If you, my reader, have not already done so, it is my earnest hope that you will pick up your broom and begin to help with daily tasks of cleaning up after the careless, the ignorant, and the greedy. Such housekeeping is especially vital where nuclear power is concerned.

One-third of the fuel rods in the reactor core of a nuclear power plant must be replaced each year; that adds up to a very large number. Because no one has figured out a safe way to permanently store the discarded rods, they are placed in "spent fuel pools" on the plant sites; these are ponds through which water is constantly pumped to carry away the heat and prevent meltdowns. This has been going on for a quarter of a century, and not only are the pools full to overflowing, there is a real fear that, added together, the rods at individual plants are enough to form critical mass and set off homegrown atomic bombs. "Well," you may be saying, "let's ship them off to a nuclear dump somewhere." But if we do, the same specter of critical mass threatens us—coupled with the real possibility of spreading the contamination.

For 25 years the Department of Energy has been trying, unsuccessfully, to solve the problem. In the early 1990s, desperate to find some place to store the nuclear waste from the 109 nuclear power plants in operation across the nation, DOE came up with a new scheme—bribery. The Office of Nuclear Waste Negotiator was set up to find a willing victim. While they would have welcomed any group—black, white, or green—who was willing to respond, someone in DOE must have reasoned that one of the Indian nations might be responsive to promises of huge amounts of money in exchange for offering its tribal lands as a "temporary," 40-year repository. (The sites will most likely become permanent, because even if the waste is removed, the land will have been permanently contaminated.) Several tribes responded.

The Sauk and Fox tribe returned the $100,000 cheque they were given for feasibility studies after Grace Thorpe (daughter of Olympic athlete Jim) and other members confronted their leaders. Other tribes, including the Chickasaw, did likewise. The Eastern Shawnee and the Miamis used the feasibility study money. After a 1993 campaign led by other experienced housekeepers (Cherokee educator Rebecca Jim, Blackfoot college student Maureen McKissick, and CASE veteran, Betty Knight Broach) they, too, turned down the DOE. (Because I'm no longer able to drive, my Eastern Shawnee granddaughter, Sandra Snelling

Patterson, took me to a meeting in Miami, where I spoke, along with
Rebecca, Grace, and June EchoHawk.)

On February 16, 1994, I scheduled a press conference to commem-
orate the twelfth anniversary of the Black Fox cancellation. When, a few
days earlier, Marla Saeger called to tell me that the Tonkawas were the latest tribe to swallow the bait, I decided to use the opportunity to remark
on the irony of CASE spending nine years preventing the production of
nuclear waste from one plant in Oklahoma, only to fall victim to the
waste from 71 other nuclear power plants.

David Rule came to the press conference to demonstrate solar devices
and brought his wife's teenaged niece, Daunena Rush. "I'm a Tonkawa,"
she told me after the conference, "and I knew nothing of these plans."
Her father and grandmother cheered her on as she wrote and sent letters
to the editors of Oklahoma and Kansas newspapers, letting her tribe—
and the general public—know what was going on.

The Tonkawa, whose name means "they stay together," are native to
Texas. In 1859 they were moved to Oklahoma, where three years later all
but a handful who had been on a buffalo hunt were massacred. Those
survivors returned once more to Texas, but after the Indian wars there,
they were moved back to Oklahoma and given 90,000 acres—all but 160
of which was taken from them in the Cherokee Strip land run.

Until 1970, their 160 acres was a pasture, and the 300 or so tribe
members lived scattered across Oklahoma and Kansas near their places of
employment. In 1970 the Bureau of Indian Affairs asked the Tonkawa to
reorganize to receive federal assistance. The government built 100 very
similar one-or two-bedroom red-brick houses, each with central heat, a
window air-conditioner, and a one-car attached garage. Most members
left their jobs for ones provided by tribal government; unemployment is
now 42%. In 1980 the Tonkawa had been given an additional 816 acres
from the lands of the disbanded Chilocco Indian School, 35 miles north
of their reservation.

It was against this background that Tonkawa President Virginia
Combrink decided to consider hosting the nuclear waste facility, accepting $200,000 for a site feasibility study with a promise of $20,000 a year
for 40 years for each Tonkawa individual. The Nuclear Waste
Negotiator's inspector decided the Tonkawa land would be fine—even
though it does not meet DOE requirements outlined in the Monitored

Retrievable Storage guidelines.

Nadine Barton has been a tireless worker in our organization. She represents CASE at Oklahoma Department of Environmental Quality hearings and other meetings and keeps a weather eye out for threats to our environment. When MRS came along, she volunteered to take me to Ponca City to give a speech at a meeting of the Kaw Nation. I ended by telling my audience that even though I didn't look Indian, my great great great grandfather was a full-blood Cherokee. It was touching to be hugged by several women and to have my hand shaken by the men as they thanked me for "stopping Black Fox."

Governor David Walters, Cherokee Chief Wilma Mankiller, Congressman Mike Synar, and Virginia Combrick's cousin, Don Patterson, were among those who came out in opposition to the MRS. Don reminded us of the many broken treaties and promises, and that the land is sacred.

Sylvia Pratt gathered several thousand signatures on petitions opposing MRS, and Marjorie Spees, Joyce Nipper, Barbara Geary, Karen Schulte, and Patrice Cuddy added their hard-won expertise. One Indian woman told Virginia Combrink, "You'll build it over my dead body!" Nothing seemed to sway her.

When Don Patterson explained to me that the media attention had died down and it was time for someone to make waves, I called a meeting in my home. About 50 people showed up, including businessman Leon Ragsdale and Sylvia Pratt, Orva Rostgeb, and Rick and Grace Klinger from the area near the proposed dump, as well as Tonkawas Richard and Marilyn Cornell. There were representatives from local garden clubs, the Audubon Society, the Sierra Club, the Oklahoma Toxics Campaign, Ecological Concerns of Oklahoma, the Umbrella, CURE, NACE and CASE. Others represented only themselves.

Every Sunday afternoon for several weekends last summer, people brought covered dishes and stayed all afternoon and into the evening, studying and strategizing ways to get the message to tribal members. The weekend that Mary Olson of NIRS came from Washington, D.C., people brought bed rolls and laid them out on my living room floor. Others set up tents under the trees.

At Lance Hughes' urging, Arjun Makhijani had invited people from tribes across the nation to an MRS seminar earlier that summer in

Washington. Grace Thorpe's cousin, Susan Williams, was convinced by what she learned there and started doing some proselytizing on her own.

We are grateful that Virginia Combrick decided to hold a referendum on August 12. Earl Hatley, Kathleen Logan, Bryan Jacobson and Chuck Ward from Oklahoma Toxics Campaign, David Rule, Roger Johnson, and Marla Saeger worked with many others to help us put together a "No Nukes" rally at a park in Tonkawa on August 7. Karen Silkwood's roommate, Sherri Ellis, came to help. Karen Cochran Moore recruited members of our group to write letters to each voting member of the Tonkawa. There were even prayer vigils—in Oklahoma City, in Tonkawa, and in Tulsa—the night before the referendum. The next day, the Tonkawas voted 58 to 44 against the dump! One more mess cleaned up.

It didn't take long for the next one to appear, and unless we act responsibly, this one will leave another ineradicable stain. On January 31, 1995, fifty-eight percent of the Mescalero Apaches voted against accepting an MRS. The thirty-three utilities that are trying to hand their problem off to someone else pressured tribal leader Chino to hold a second referendum. Despite the hard work of Rufina Laws and Joe Geronimo, the vote was reversed.

In January, U.S. Senator Bennett Johnston[2] introduced Senate bill 167 (Nuclear Waste Policy Act), which would force 43 states to accept the risk from 70,000 metric tons of spent nuclear fuel rods traversing their highways and railroads over the next 30 years and would permit the contamination of groundwater by the waste. An active fault runs through Yucca Mountain,[3] where the Department of Energy is now spending $2,000[4] per linear inch to drill deep tunnels to receive this waste. Even if there were never another earthquake, this would be a lethal decision. There is strong evidence they will not be able to store the waste there without causing a nuclear explosion that would blow the top off the mountain.[5]

Even if every truck and every train to the site were to arrive safely with its cargo intact (and DOE itself expects there would be 15 accidents a year on the average) and even if the faults were quiet for hundreds of millennia, this would not be a permanent solution. On March 5, 1995, The *New York Times* reported the warnings of two Los Alamos National Laboratory scientists: Plutonium stored at Yucca Mountain would remain long after the steel casks holding it dissolved. One of the scientists, Dr. Charles Bowman, said that at that point, the plutonium could migrate and concentrate, while the rock containing it could actually accelerate a chain reaction and subsequent explosion, blowing the top off the mountain.

Instead of ending the production of high-level radioactive waste, the nuclear industry is planning to continue the operation of the aging plants and is mounting a new campaign to build more plants. There is no plan for ending the production of this waste or for managing it. Instead, the waste will most likely be stored above ground where it is vulnerable to sabotage as well as to the forces of nature and radioactive decay. This is a formula for putting this lethal waste out of sight and hoping that it will then be out of mind—for sweeping it under the rug. But as every housekeeper knows, what is swept under the rug eventually filters out again.

Once more, I have put aside my need to earn a livelihood to help organize a campaign to defeat this bill and others that have also been introduced. From a three-day national conference on this issue in Las Vegas in February, 1995, Nadine Barton, Sylvia Pratt, Grace Thorpe, and Doris Gunn brought back information and ideas enough to keep us all busy. The DOE attempted to lull the public into ignoring the problems, but eighty percent of Nevadans see this bill as a disaster in itself.

With the breakup of the former Soviet Union, more housekeeping became a necessity. Soviet experiences illustrate why we must not lose sight of the importance of human error, unpredictable economics and politics, and global contamination as we strive to keep up with nuclear housework.

On April 26, 1986,[6] engineers at a nuclear plant in Chernobyl, near Kiev, were conducting a test of their emergency systems. They reduced power to below twenty-five percent in one reactor, triggering an explosion and meltdown of the reactor core, fatally irradiating two operators and thirty other workers and contaminating about 30,000 square miles of farmland—in Europe as well as the then-Soviet Union. Many billions of dollars were lost in farm productivity, Laplanders had to dispose of reindeer that they depended upon for their livelihood, and the radiation went around the world—even across Oklahoma! A human being had neglected to warn workers that low power would make the graphite-core reactor dangerously unstable.

Late last summer—eight years after the meltdown—Patricia was in Petersburg, nearly a thousand miles north of Chernobyl. She found the local newspapers still full of articles warning people about eating the mushrooms they traditionally gather from the forests that time of year. Radiation monitoring equipment is routinely available in Petersburg for those who fear that farm produce has been contaminated by the soil in which it was grown. When she and her party ordered bottled water (because of the danger of *giardia* in the local water supply), their taxi driver thought they were afraid of radioactive water; he shrugged and said that you had to learn to live with it: It was everywhere.

The melted reactor has been encased in an inadequate, unsafe "sarcophagus" because Ukraine does not have enough money to build a minimally sound one. As it is, fifteen percent of Ukraine's budget is spent on Chernobyl and its consequences—including early retirement for some 750,000 people involved in the cleanup, whose lives are expected to be shortened drastically. (About 30,000 are already disabled, and many have already died.) The water supply for the great city of Kiev is contaminated, and there is no money for minimizing the contamination—if, indeed, it is even possible. Thyroid cancer in children, previously almost unheard of, is now endemic in the area (200 times normal, according to the *British Medical Journal*) and the incidence of deaths from other cancers and respiratory disorders has risen dramatically. Birth defects are heartrending.

Fire shut down a second reactor at the Chernobyl power complex in 1991, but reliance on nuclear energy had allowed research and investment in other sources to languish and has forced the Ukraine to continue operating the remaining two reactors. During the spring of 1994 the Ukranian government announced it was determined to shut down Chernobyl but could not do so until other ways were found to meet the national energy shortage. For the same reason, fifteen other reactors condemned as unsafe are still operating in Ukraine, Russia, and Lithuania.

Authorities in Kazakhstan, however, had the statesmanship to enlist the help of President Clinton in protecting the world from nuclear warheads stored in their country. They knew that Balkan political instability made it impossible for them to be sure they could carry out their responsibility, so they found someone to share it.

We cannot congratulate ourselves that superior engineering and greater financial resources make our own nuclear reactors safer. When

TMI-2 and Browns Ferry nearly melted down, they were new reactors; they were damaged through human error. Today, we are surrounded by aging reactors whose containment vessels and other parts have been made increasingly brittle by radiation. Containment vessels have begun cracking; they will not necessarily begin leaking—and thus give warning—before they break.[7] Nuclear fuel rods are failing in increasing numbers.[8] Seven of our currently operating boiling water reactors have developed cracks in their core shrouds.[9] All these problems and others are potential causes of meltdowns.

Renewable energy technologies and energy-conserving appliances are no longer exotic or especially expensive. Wind power is already cheaper than nuclear or coal-produced energy; within the next five years, its cost will drop even lower—to around 4 cents per kwh. With the new technology available, solar energy now costs only about 6 cents. Hydrogen is not only clean and, properly produced, inexpensive, it is an ideal way to save energy produced in low-demand periods for use in times of high demand.

Industrial giants that have been at the forefront of technological improvement for decades are joining companies begun specifically to manufacture and promote alternative-energy equipment. Because much of this technology is accessible to individual homeowners and gives them the opportunity to become free of the utility monopolies, I have included some of these sources in the bibliography at the end of this book.

In contrast to the increasing cost-accessibility of alternative sources of energy, costs of fossil and nuclear fuels will, at best, remain constant and are more likely to increase.[11]

In recent years three commercial reactors—Yankee Rowe in Massachusetts, San Onofre-1 in California, and Trojan in Oregon—have been permanently shut down. There is hope that before this book is published, three TVA plants will have bitten the dust. This is a start.

In 1987, former Nuclear Regulatory Commissioner Asselstine predicted that within the next ten to twenty years there would be a Chernobyl-type accident in the United States. We cannot afford to wait for our own Chernobyl or our own Chelyabinsk to rid ourselves of this lethal experimental technology. Before it is too late, we must follow through and eliminate all nuclear power plants.

Each of us bears individual responsibility in the task of putting our house in order, but we can be more effective working together. It is my earnest hope that each of my readers will become members of organizations, local and national, and will work in love and harmony for all our futures. Many hands not only make light work, they make lighter spirits.

Notes

Chapter 4 – Enlisting Robert

[1]R. Halasz (ed.), *Illustrated Encyclopedia Yearbook 1975*. Chicago: Rockville, 1975.

[2]*Bulletin of the Atomic Scientists*. October 1974.

[3]Sierra Club *Bulletin*. January 1974.

[4]Ibid.

[5]December 5, 1974.

[6]Consumer Action Now (CAN). New York.

[7]R. Halasz, *op. cit.*

Chapter 7 – "God Made Rattlesnakes, too."

[1] Wednesday, September 5, 1973.

Chapter 9 – Our Paul Revere

[1]A film, *Lovejoy's Nuclear War*, is available from Green Mountain Post Films, Montague, MA.

Chapter 10 – "And you work for PSO?"

[1]Jerry Stoll, Director. New York: Impact Films, 1973.

Chapter 11 – A Hill of Beans

[1]J.I. Rodale, a tireless worker for health throughout the world. In J.I. Rodale's lexicon, world peace and healthy environment were necessities for health.

Chapter 12 – Camp Gruber

[1]Wilma Mankiller and Michael Wallis. *Mankiller: A Chief and Her People*. New York: St. Martin's Press, 1993. Pp. 62-3.

[2]March 25, 1974.

[3]Dean Abrahamson, a physician and physicist, who was then Associate Professor of Public Affairs and Director of the Center for Studies of the Physical Environment at the University of Minnesota.

[4] *Honecker vs. Hendrie*. 156 Drakes Lane, Summertown, TN 38483: The Book Publishing Co., 1978.

Chapter 13 – The Debate Continues: 1974-75
[1]Letters to the Editor. *The Saturday Review.* June 15, 1974.
[2]Letters to the Editor. *The Saturday Review.* September 7, 1974.
[3]Nuclear Fuel Services in West Valley, New York.

Chapter 14 – Critical Mass '74 and Karen Silkwood
[1] Howard Kohn, *The Killing of Karen Silkwood,* Summit Books, NY, 1981.

Chapter 16 – "Don't tell *me* I can't!"
[1] The Farm, 156 Drakes Lane, Summertown, TN 38483
[2]Jeannine Honicker, *Honicker vs. Hendrie.* 156 Drakes Lane, Summertown, TN 38483. The Book Publishing Co.

Chapter 20 – Delay, Delay, Delay
[1]The hearing was held at the U.S. Courthouse at Denver and 4th Streets in downtown Tulsa before a licensing board consisting of Chairman Daniel M. Head and members Walton Purdom and Frederick Shon with their lawyers, Dow Davis and James Tourtellotte. PSO sent attorneys, Michael Miller and Paul Murphy from the Chicago law firm of Isham, Lincoln & Beale, and their local lawyer, Charles Crane. Tom (Andrew T.) Dalton represented CASE and Ilene Younghein.
[2]Bob Mycue, "Splitting the Atom Split the People." *Tulsa World,* September 6, 1976.

Chapter 21 – Six More Months of Prehearing Conferences
[1]Convened at 10 a.m. on Thursday, January 6, 1977, in the U.S. Courthouse in downtown Tulsa.

Chapter 23 – Martha and Mary
[1]August 14, 1977.
[2]Convened at 1 p.m., Monday, August 22, 1977, in the main conference room of the Army Corps of Engineers district offices at Third Street and Boulder Avenue in downtown Tulsa by NRC attorney Sheldon J. Wolfe, chairman of the three-member Atomic Safety and Licensing Board. Other board members were NRC physicist Frederick J. Shon and Dr. Paul Purdom, director of the Institute for Environmental Studies.

[3]Public Service Company of Oklahoma, *Black Fox Station Application for Licenses*, Construction Permit Stage, 1977.

[4]Public Service Company of Oklahoma, *Preliminary Safety Analysis Report, Black Fox Station*, 1977.

[5]Public Service Company of Oklahoma, *Environmental Report, Black Fox Station*, 1977.

[6]Jack Anderson with Les Whitten. *Oklahoma Journal*, November 2, 1977.

[7]Joseph E. Howell, "3 women stand against PSO at hearings on nuclear plant." *Tulsa Tribune*, Thursday, August 25, 1977, Section d, 1-2.

Chapter 24 – How to Kill a Turkey

[1]Paul J. Cleary, "John Zelnick Enters Senate Race; Details Election Platform." *Tulsa World*, November 18, 1979.

Chapter 25 – Tom Dalton's Bombshell

[1] Joseph E. Howell, "Sun brightens PSO's future." *Tulsa Tribune*, April 25, 1980.

[2] Until Friday, October 21, 1977.

Chapter 26 – The Dollar Value of a Human Life

[1] *Tulsa World*, May 5, 1978.

Chapter 33 – Parting the Red Sea

[1]Tuesday, October 10, 1978, at the Camelot Inn at South Peoria and Highway I-44. Reactor safety, fire protection, the reactor's control system and design and operational aspects of the GE reactor were issues to be covered. Another issue admitted for the hearing was the danger of explosions of hydrogen escaping from the off-gas system in boiling water reactors.

Staff members representing the NRC were Dow Davis, William Paton and Colleen P. Woodhead. Attorneys on behalf of PSO were Joseph Gallo, Paul M. Murphy and Glenn Nelson, from the law firm of Isham, Lincoln and Beale.

Chapter 34 – "David vs. Goliath"

[1] *WASH-1400*.

[2] *Tulsa Tribune*, Thursday, October 12, 1978.

Chapter 35 – "A Summary Would Not Serve the Purpose."
[1] Joe Howell, Tulsa Tribune, Friday, October 13, 1978.

Chapter 36 – The Reed Report
[1] Tuesday, December 5, 1978.
[2] Gregory Minor, Palo Alto, California, partner in MHB Technical Associates, and CASE's expert witness.
[3] In the Federal Building in downtown Tulsa.

Chapter 37 – "The Banks Won't Finance Them."
[1] Excerpts from *Fragment of a Scattered Rainbow*, 1987, available from the author.
[2] "Nader to speak at Lloyd Noble, Wednesday, August 27, 1980." *The Oklahoma Daily*, Norman, Oklahoma.

Chapter 40 – Black Fox and Three Mile Island
[1] February 19-20, 1979, in the Page Belcher Federal Building in downtown Tulsa.
[2] Wednesday, March 28, 1979.
[3] According to James Broughton of the Idaho National Engineering Laboratory in a November, 1988, report.
[4] "Letter to Richard O. Newman." *Tulsa World*, April 1, 1979.
[5] "STOP BLACK FOX." *Tulsa Tribune*, April 11, 1979.
[6] Richard Wheatley, "300 Jam Hearing in BA on N-Plants." *Tulsa World*, April 20, 1979.
[7] April 25, 1979.
[8] Chuck Ervin, "Cummins Sinks Own Nuclear Moratorium Resolution." *Tulsa World*, April 74.
[9] Mark Blackburn, "Wall Street Snubs Nukes." *The Oklahoma Observer*, July 10, 1979.
[10] "City to be evacuated in Black Fox disaster." *Broken Arrow Ledger*, January 11, 1980.
[11] *Energy Daily*, June 22, 1979, and *Nucleonics Week*, June 14, 1979.

Chapter 41 – The Lure of Easy Money
[1] Bob Mycue, "PSC Files $29.4 Million Rate Request." *Tulsa World*, September 5, 1979.

[2]Allison, Gary D. "Judging the Prudence of Constructing Nuclear Power Plants: A Report to the Oklahoma Corporation Commission." *University of Tulsa Law Journal*, 1979, 15(2), 262-298.

[3]"Claremore musicians play in CASE concert." *Claremore Progress*, Friday, September 12, 1980.

[4]"Danger: House Bill 1207." *Soft Path* (Claremore, Oklahoma), March 1981.

Chapter 43 – Victory Is Ours!

[1]September 14, 1981.

Chapter 44 – The Victory Celebration

[1]Kelly Jennings, "Victory at Last ... Black Fox is Dead!" *The Oklahoma Sierran*, 1982.

[2]Stefan Kanfer, "They Did the Right Thing." *People*, Fall 1989 (extra), 118-9.

Epilogue – Keeping House

[1] Including, arsenic, barium, cadmium, mercury, lead, molybdenum, selenium, vanadium, and zinc.

[2]D. Louisiana

[3]James Coates, "A jolt to a nuclear plan." *Boston Globe*, July 29, 1991.

[4]Tim Beardsley, "Pass the Plutonium, Please." *Scientific American*, March, p.18-29.

[5]Helen Caldicott, Nuclear Madness, *W.W. Norton & Co.*, New York, N.Y.

[6]Chernobyl Meltdown, *Newsweek*, May 12, 1986. p. 20-49.

[7]*Greenpeace*, March 26, 1993.

[8]Ashley Craddock, "Faulty rods." *Mother Jones*, May-June, 1994, p. 24.

[9]*The Nuclear Monitor* (NIRS), November 21, 1994.

[10]Ibid

[11]"Company spotlight: New world power." *Business*, 1994, 16(3), 48.9.

Glossary

ad hominem Attacks (usually in debate) on the person rather than on what the person says

A&M Oklahoma Agricultural and Mechanical College, now OSU

ACRS Advisory Committee on Nuclear Safeguards

AEC Atomic Energy Commission (pre-1974)

ASLB Atomic Safety and Licensing Board

ATWS Anticipated Transients Without Scram. Situations in which it may be expected that something will go wrong, but in which the reactor does not shut down automatically.

CASE Citizens' Action for Safe Energy, Inc., successor to Citizens' Action Group

CEQ [White House] Council on Environmental Quality

core shrouds Water pipes that direct the flow of radioactive water around the fuel rods to ensure adequate cooling

CWIP Construction Work in Progress. Funds to pay for new construction that are added to the utility rate base rather than subtracted from stockholders' dividends

DOE The federal Department of Energy

ECCS Emergency Core Cooling System

EIS Environmental Impact Study

eminent domain A court proceeding based on the doctrine that the people have ultimate power (eminent domain) over all land and that the good of the whole overrides the good of the individual. While this is normally used for landtaking for government projects, the utilities have historically been treated by the courts as quasi-government entities for this purpose.

EPA Environmental Protection Agency

ERDA Energy Research and Development Administration

FES Final Environmental Statement

FOE Friends of the Earth

fuel enrichment plant A manufacturing facility where the proportion of U^{235} in reactor fuel is increased from 2% to 4% to make the fuel sufficiently radioactive to run the reactor efficiently.

fuel fabrication plant   A facility where enriched uranium is made into pellets and loaded into stainless steel or zircalloy tubes 12 feet long and 1/2 inch in diameter. The tubes are capped and welded shut before being assembled in groups of several hundred for shipping. Each current reactor requires 50,000 such fuel rods.

fuel reprocessing plant   A facility in which the used-up fuel rods are chopped up and dissolved in nitric acid. When such plants were in operation, the uranium, plutonium, and other radioactive salts that resulted were then changed into other physical and chemical forms for disposal or further use. The plant at West Valley, New York, polluted its surroundings so severely that its owner, Nuclear Fuel Services, could not continue to operate it and surrendered it to the New York State government. GE plants in Illinois and Barnwell, South Carolina, are also shut down, leaving no civilian reprocessing plants currently in service.

GAO   General Accounting Office. A federal watchdog agency.

heliostat   A mirror designed to track the sun as it passes overhead and to reflect the energy of the sun's rays to a heat receiver on top of a tower in the center of the field.

in camera   Literally, "in [the judge's] chambers," a phrase used for proceedings that are not to become public knowledge.

JCAE   Congressional Joint Committee on Atomic Energy. Created by statute in 1946, the Joint Committee was made up of members of both houses of Congress and charged with the responsibility for examining all bills that dealt primarily with nuclear power.

LPN   Licensed Practical Nurse

MIT   Massachusetts Institute of Technology

MRS   Monitored Retrievable Storage

NACE   Native Americans for a Clean Environment

NASA   National Aeronautical and Space Administration

NEPA   National Environmental Policy Act

NIRS   Nuclear Information and Resource Service

NPDES   Nuclear Pollution Discharge Elimination System

NRC   Nuclear Regulatory Commission

NRDC   National Resources Defense Council

OAMC   Oklahoma Agricultural and Mechanical College, now OSU

OCC   Oklahoma Corporation Commission, the organization charged with regulating utilities in Oklahoma

OPEC Oil-Producing Export Countries

OSU Oklahoma State University at Stillwater

OU The University of Oklahoma at Norman

OWRB Oklahoma Water Resources Board

pro bono Actually, *pro bono publi-co,* a wonderful phrase that means, literally, "for the public good" or "for the good of the people." In practice, it means free of charge.

PSC or PSO Public Service Company of Oklahoma

PSAR Preliminary Safety Analysis Report

raffinate A word invented to indi-cate a salt of a product or products of a refining process.

RN Registered Nurse

RSS Rasmussen Report, *q.v.*

photovoltaic solar cells Treated sil-icon semiconductors that convert solar radiation to electricity. Gram for gram over their respective life-times, thin-film amorphous silicon cells produce as much energy as ura-nium in a nuclear reactor. Silicon, a major constituent of sand, is an envi-ronmentally friendly material.

scram "anticipated transients with-out scram" Scram is the sudden shutdown of a nuclear reactor, usu-ally by rapid insertion of safety rods. Emergencies that may come and go are transient, that is, not chronic or permanent, but they are also anticipated, because they are expected to happen at odd inter-vals. When they occur, the reactor operator or automatic control equipment is supposed to *scram* the reactor. *Anticipated Transients Without Scram* are events—like, for instance, power surges—that cause an accident and at the same time make it impossible to shut down the reactor.

SFC Sequoyah Fuels Corporation

TVA Tennessee Valley Authority

UCS Union of Concerned Scientists

Resources

For those who wish to pursue the subjects of nuclear energy and weapons further and to explore alternative energy possibilities, I offer a minimal list of resources and source materials. The perusal of any of the monographs and periodicals below will provide additional sources. The organizations on the list that follows have not only been cooperative and reliable resources to me on any conceivable subject related to nuclear and/or alternative energy issues, they offer avenues for cooperative citizen action.

Monographs

Bertell, Rosalie. *No immediate danger.* Summertown, Tennessee: The Book Publishing Company, 1982.

Brower, Michael. *Cool energy: Renewable solutions to environmental problems.* Cambridge: MIT Press, 1992.

Caldicott, Helen, M.D. *Nuclear madness.* New York: Norton, 1994.

Clausen, Peter A. *Nonproliferation and the national interest: America's response to the spread of nuclear weapons.* New York: Harper Collins, 1992.

Davidson, Joel. *The new solar electric home: The photovoltaics how-to handbook.* Ann Arbor, Michigan: Aztec, 1987. (P.O. Box 7119, zip code 48107, phone 313-995-1490)

Franck, Irene, and David Brownstone. *The Green Encyclopedia.* New York: Prentice Hall, 1992.

Georgy, Anna, and Friends. *No Nukes: Everyone's guide to nuclear power.* Boston: South End, 1979.

Gilgartner, Stephen. *Nukespeak: Nuclear language, visions, and mindset.* San Francisco: Sierra Club Books, 1982.

Gipe, Paul. *Wind power for home and business: Renewable energy for the 1990s and beyond.* Post Mills, Vermont: Chelsea Green, 1993. (P.O. Box 130, Route 113, zip code 05058-0130)

Gofman, John W., and Arthur Tamplin. *Poisoned power: The case against nuclear power plants.* Emmaus, Pennsylvania: Rodale, 1972.

Honicker, Jeannine. *Honicker vs. Hendrie: A lawsuit to end atomic power.* Summertown, Tennessee: The Book Publishing Company, 1978.

Kerr, Robert S. *Land, wood and water.* Fleet Publishing Corporation, 1960.

Kohn, Howard. *Who killed Karen Silkwood?* New York: Summit, 1981.

Makhijani, Arjun, and Saleska Scott. *High-level dollars, low-level sense.* New York: Apex, 1992.

Mankiller, Wilma, and Michael Wallis. *Mankiller: A chief and her people.* New York: St. Martin's, 1992.

Medvedev, Zhores. *A nuclear disaster in the Urals.* New York: Norton, 1979.

Miner, Craig. *Wolf Creek Station.* Ohio: Ohio State University Press, 1993.

Peavey, Michael A.. *Fuel from water: Energy independence with hydrogen.* Louisville, Kentucky: Merit Products Inc., 1993.

Strong, Steven J., with William G. Scheller. *The solar electric house: A design manual for photovoltaic power systems.* Emmaus, Pennsylvania: Rodale, 1987.

Tohl, Frederik. *Chernobyl.* New York: Bantam Publishing, 1988.

Wasserman, Harvey, Norman Solomon, Robert Alvarez, and Eleanor Walters. *Killing our own: The disaster of America's experience with atomic radiation.* New York: Delacorte, 1982.

Periodicals

Bulletin of the Atomic Scientists. Published 10 times a year by the Educational Foundation for Nuclear Science, 1020-24 East 58th Street, Chicago, Illinois 60637.

Calypso Log. Published quarterly by the Cousteau Society, Inc., 777 Third Avenue, New York, New York 10017.

The Energy Monitor: Citizen Watchdog of the TVA. Published by the Tennessee Valley Energy Reform Coalition, P.O. Box 8290,

Knoxville, Tennessee 37996. A new and extremely informative publication.

Greenpeace. Published bimonthly by Greenpeace USA, 1436 U Street, NW, Washington, D.C. 20009.

*Home Power Magazine: The hands-on journal of home made power.** Published bimonthly. P.O. Box 520, Ashland, OR 97520. A good source of information on many kinds of renewable and alternative energy sources and their applications.

Mother Jones. Published monthly by the Foundation for National Progress, 625 Third Street, San Francisco, California 94107.

EnviroAction. Published regularly when Congress is in session to inform members of the National Wildlife Federation about legislative action on environmental issues. The publication is available free, upon request, to NFW members. For More Information: Office of Grassroots Action, 1400 16th St., N.W., Washington, DC 20036—2266.

Not Man Apart. Published monthly by Friends of the Earth, 529 Commercial Street, San Francisco, California 94111.

The Nuclear Monitor. Published biweekly by the Nuclear Information and Resource Service, 1424 16th Street NW, Suite 601, Washington, D.C. 20036.

Nucleus. Published quarterly by the Union of Concerned Scientists, 26 Church Street, Cambridge, Massachusetts 02238.

Prevention. Published monthly by Rodale Press, Inc., Emmaus, Pennsylvania 18049.

Public Citizen. Published monthly by Public Citizen, Inc., P.O. Box 7229, Baltimore, Maryland 21218.

Rachel's Hazardous Waste News. Published weekly by Environmental Research Foundation, P.O. Box 73700, Washington, D.C. 10056-3700.

*Real Goods Catalog.** Published four to six times a year at 966 Mazzoni Street, Ukiah, CA 95482-3471. 300-762-7325. A catalogue that reads like a magazine, giving quite detailed and comparative information on alternative energy technology. It carries equipment from industrial giants and cottage industries, and offers valuable suggestions on their applications.

Rocky Mountain Institute Newsletter. Published quarterly by the Rocky Mountain Institute (founded 1976 by Amory Lovins), P.O. Drawer 248, Old Snowmass, Colorado 81564. (303) 927-3851 A clearinghouse for the intelligent use of energy.

Sierra. Published bimonthly by the Sierra Club, 730 Polk Street, San Francisco, California 94109.

*Solar Today.** Published bimonthly by the American Solar Energy Society, 2400 Central Avenue, G-1, Boulder, CO 80301-2843. Information on the latest solar products and services, showcasing the fact that the solar industry is alive and well and ready to do business in today's marketplace.

* *Sources of information on buying the latest in alternative energy technology.*

Organizations

Citizens' Action for Safe Energy, Inc. (CASE)
3609 East Blue Starr Drive,
Claremore, Oklahoma 74017. (918) 341-3494

Citizens' Clearinghouse for Hazardous Wastes (CCHW)
The Center for Environmental Justice,
P.O. Box 6806, Falls Church Virginia 11040.
(703)237-2249
Founded by Lois Gibbs in the wake of the Love Canal closing.

Greenpeace USA
1436 U Street, NW,
Washington, D.C. 20009.
(202) 462-1177

This colorful, but extremely dedicated international group of environmentalists and peace activists is very successful at drawing attention to issues of world concern.

Nuclear Information and Resource Service (NIRS)
1424 16th Street NW, Suite 601,
Washington, D.C. 20036.
(202) 328-0002

An information clearing house that provides up-to-date information on congressional legislation, citizen initiatives, and nuclear power plants across the nation.

Oklahoma Toxics Campaign
P.O. Box 74
Guthrie, Oklahoma 73044

A core remnant of the National Toxics Campaign and the largest environmental organization in Oklahoma, this very active and dedicated group makes its expertise and experience available to groups inside and outside the state.

Public Citizen CRITICAL MASS Energy Project
215 Pennsylvania Avenue, SE,
Washington, DC 20003
(202) 547-7392.

Founded in 1974 by Ralph Nader, this organization has been a powerful voice in the movement to decrease reliance on nuclear and fossil fuels and to promote safe, economical, and environmentally sound energy alternatives. Working closely with other citizens' groups and individuals across the country, CRITICAL MASS prepares and disseminates reports, lobbies Congress and serves as a watchdog for key federal and state regulatory agencies.

Sierra Club
730 Polk Street,
San Francisco, CA 94109.
(415) 776-2211

The oldest, most influential environmental organization in the world, the Sierra Club has a chapter in every state.

Union of Concerned Scientists
26 Church Street,
Cambridge, Massachusetts 02238.
(617) 547-5552

Very involved in renewable energy advocacy as well as preventing nuclear proliferation, this group provides teachers with kits for instruction modules on renewable energy and is a respected resource to the general public.

Index

BLACK FOX